Python for Beginners

Python is an amazing programming language. Getting started with Python is like learning any new skill. It is a powerful programming language that can be used to develop any specific application, right from the social media apps like Pinterest and Instagram, to research programming interfaces for top institutes like NASA, Google, and other defense and research organizations. Beginners can easily grasp the language, and is comparatively easier than R and MATLAB. It has a relatively simple syntax and coding rules, making it easy to learn for beginners.

This book starts out with a walkthrough of the basic Python elements and data structures, working through variables, strings, numbers, lists, and tuples, outlining how you work with each of them. Next, if statements and logical tests are covered, followed by a dive into dictionaries. After that, the book covers user input, while loops, functions, classes, and file handling, as well as code testing and debugging. With the abundance of resources, it can be difficult to identify which book would be best for your situation.

Main Features of this Book:

- It is clean and elegant coding style
- It is suitable for beginners who have no knowledge about programming
- It has very simple sequence sets of strings
- This book has feature to draw animated picture, such as Tkinter
- It covers Data Structure and Database Connectivity in Python
- It also Covers CGI Gateway and Networking in Python

This book focuses on enthusiastic research aspirants who work on scripting languages for automating the modules and tools, development of web applications, handling big data, complex calculations, workflow creation, rapid prototyping, and other software development purposes. It also targets Graduates, Post graduates in Computer Science, Information Technology, Academicians, Practitioners, and Research Scholars.

Python for Beginners

Kuldeep Singh Kaswan
Jagjit Singh Dhatterwal
B. Balamurugan

CRC Press
Taylor & Francis Group
Boca Raton London New York

CRC Press is an imprint of the
Taylor & Francis Group, an **informa** business

A CHAPMAN & HALL BOOK

First edition published 2023
by CRC Press
6000 Broken Sound Parkway NW, Suite 300, Boca Raton, FL 33487-2742

and by CRC Press
2 Park Square, Milton Park, Abingdon, Oxon, OX14 4RN

Library of Congress Cataloging-in-Publication Data
Names: Kaswan, Kuldeep Singh, author. | Dhatterwal, Jagjit Singh, author. |
Balamurugan, B, author.
Title: Python for beginners / Kuldeep Singh Kaswan, Jagjit Singh Dhatterwal, B Balamurugan.
Description: First edition. | Boca Raton : Chapman & Hall/CRC Press, [2022] | Includes bibliographical references and index. | Summary: "Python is a powerful programming language that can be used to develop any specific application, right from the social media apps like Pinterest and Instagram, to research programming interfaces. Beginners can easily grasp the language as it has a relatively simple syntax and coding rules. This book starts out with a walkthrough of the basic Python elements and data structures, working through variables, strings, numbers, lists, and tuples, outlining how you work with each of them. It covers statements and logical tests followed by a dive into dictionaries along with user input, while loops, functions, classes, and file handling, as well as code testing and debugging"-- Provided by publisher.
Identifiers: LCCN 2021027896 (print) | LCCN 2021027897 (ebook) | ISBN 9781032063867 (hbk) | ISBN 9781032063881 (pbk) | ISBN 9781003202035 (ebk)
Subjects: LCSH: Python (Computer program language) | Computer programming.
Classification: LCC QA76.73.P98 K37 2022 (print) | LCC QA76.73.P98 (ebook) | DDC 005.13/3--dc23
LC record available at https://lccn.loc.gov/2021027896
LC ebook record available at https://lccn.loc.gov/2021027897

ISBN: 978-1-032-06386-7 (hbk)
ISBN: 978-1-032-06388-1 (pbk)
ISBN: 978-1-003-20203-5 (ebk)

DOI: 10.1201/9781003202035

Typeset in Palatino
by SPi Technologies India Pvt Ltd (Straive)

Contents

Preface

This book assumes that you are a new programmer with no prior knowledge of programming. So, what is programming? Programming solves problems by creating solutions – writing programs – in a programming language. The fundamentals of problem solving and programming are the same regardless of which programming language you use. You can learn programming using any high-level programming language such as Python, Java, C++, or C#. Once you know how to program in one language, it is easy to pick up other languages, because the basic techniques for writing programs are the same. So what are the benefits of learning programming using Python? Python is easy to learn and fun to program. Python code is simple, short, readable, intuitive and powerful, and thus it is effective for introducing computing and problem solving to beginners.

We use the Python programming language for all of the programs in this book – we refer to "Python" after "programming" in the title to emphasize the idea that the book is about *fundamental concepts in programming*. This book teaches basic skills for computational problem solving that are applicable in many modern computing environments; it is a self-contained treatment intended for people with no previous experience in programming.

This book takes you step-by-step through the process of learning the Python programming language. Each line of the code is marked with numbers and is explained in detail. All the names of variables, strings, lists, dictionaries, tuples, functions, methods and classes consist of several natural words, and in the explanation part they are provided in a different font to indicate to readers that they are part of programming code and to distinguish them from normal words. This programming style of using readable natural names makes the reading of code a lot easier and prevents programming errors.

We teach problem solving in a problem-driven way and do not focus on syntax. We stimulate student interest in programming by using interesting examples in a broad context. While the central thread of the book is on problem solving, the appropriate Python syntax and library are introduced. To support the teaching of programming in a problem-driven way, we provide a wide variety of problems at various levels of difficulty to motivate students. In order to appeal to students in all majors, the problems cover many application areas in math, science, business, financial management, gaming, animation and multimedia.

The best way to teach programming is *by example*, and the only way to learn programming is *by doing*. Basic concepts are explained by example and a large number of exercises with various levels of difficulty are provided for students to practice.

Editor Biographies

Dr. Kuldeep Singh Kaswan is presently working as Professor at the School of Computing Science & Engineering, Galgotias University, Uttar Pradesh. He received a doctorate in Computer Science at the faculty of Computer Science at Banasthali Vidyapith, Rajasthan. He received a Master of Technology in Computer Science and Engineering from Choudhary Devi Lal University, Sirsa (Haryana). His area of interest includes software reliability, soft computing and machine learning. He has published a number of research papers, books, book chapters and patents at the national and international level. He can be reached by e-mail at: kaswankuldeep@gmail.com.

Dr. Jagjit Singh Dhatterwal is presently working as Associate Professor at the Department of Artifical Intelligence & Data Science, Koneru Lakshmaiah Education Foundation, Andhra Pradesh. He received a Master of Computer Application from Maharshi Dayanand University, Rohtak (Haryana). He is also a member of the Computer Science Teacher Association, the International Association of Engineers, the International Association of Computer Science and Information Technology, the Association of Computing Machinery and of the Computer Society of India. His area of interest includes artificial intelligence and multi-agent technology. He has a number of publications in international and national journals and has presented at conferences.

Dr. B. Balamurugan is currently working as Professor in the School of Computing Sciences and Engineering at Galgotias University, Greater Noida, India. His contributions focus on Engineering Education, Blockchains and Data Sciences. He has 12 years experience working as a faculty member at VIT University, Vellore. He has published high impact papers for the IEEE. He has worked on more than 50 edited and coauthored books and collaborated with eminent professors across the world. He completed his bachelor's, master's and PhD degrees in India. His passion is teaching and adapting different design thinking principles while delivering his lectures. He serves on the advisory committee for several startups and forums and does consultancy work for industry on the Industrial IOT. He has given over 175 talks at various events and symposia.

1

Introduction to Python

1.1 Introduction

Python is a free general-purpose programming language with beautiful syntax. It is available across many platforms including Windows, Linux and macOS. Due to its inherently easy to learn nature along with object oriented features, Python is used to develop and demonstrate applications quickly. It has the "batteries included" philosophy wherein the standard programming language comes with a rich set of built-in libraries. It's a known fact that developers spend most of their time reading code rather than writing it and Python can speed up software development. Hosting solutions for Python applications are also very cheap. The Python Software Foundation (PSF) nurtures the growth of the Python programming language. A versatile language like Python can be used not only to write simple scripts for handling file operations but also to develop massively trafficked websites for corporate IT organizations [1].

1.2 Software

According to IBM Research: "Software development refers to a set of computer science activities dedicated to the process of creating, designing, deploying and supporting software." Software itself is the set of instructions or programs that tell a computer what to do. It is independent of the hardware and makes computers programmable. There are three basic types:

- **System software** to provide core functions such as operating systems, disk management, utilities, hardware management and other operational necessities.
- **Programming software** to give programmers tools such as text editors, compilers, linkers, debuggers and other tools to create code.

DOI: 10.1201/9781003202035-1

- **Application software** (applications or apps) to help users perform tasks. Office productivity suites, data management software, media players and security programs are examples. Applications also refers to web and mobile applications like those used to shop on Amazon, to socialize using Facebook or to post pictures to Instagram.

A possible fourth type is **embedded software**. Embedded systems software is used to control machines and devices not typically considered to be computers, such as telecommunications networks, cars and industrial robots. These devices, and their software, can be connected as part of the Internet of Things (IoT).

Software development is primarily done by programmers, software engineers and software developers. These roles interact and overlap, and the dynamics between them vary greatly across development departments and communities.

- **Programmers, or coders**, write source code to program computers for specific tasks like merging databases, processing online orders, routing communications, conducting searches or displaying text and graphics. Programmers typically interpret instructions from software developers and engineers and use programming languages like C++ or Java to carry them out.
- **Software engineers** apply engineering principles to build software and systems to solve problems. They use a modeling language and other tools to devise solutions that can often be applied to problems in a general way, as opposed to merely solving for a specific instance or client. Software engineering solutions adhere to the scientific method and must work in the real world, as with bridges or elevators. Their responsibility has grown as products have become increasingly more intelligent with the addition of microprocessors, sensors and software. Not only do more products rely on software for market differentiation, but their software development must be coordinated with the product's mechanical and electrical development work.
- **Software developers** have a less formal role than engineers and can be closely involved with specific project areas, including writing code. At the same time, they drive the overall software development lifecycle, including working across functional teams to transform requirements into features, managing development teams and processes, and conducting software testing and maintenance.

The work of software development isn't confined to coders or development teams. Professionals such as scientists, device fabricators and hardware makers also create software code even though they are not primarily software developers. Nor is it confined to traditional information technology industries such as software or semiconductor businesses. In fact, according to the Brookings Institute, those businesses "account for less than half of the companies performing software development."

An important distinction needs to be made between custom software development and commercial software development. Custom software development is the process of designing, creating, deploying and maintaining software for a specific set of users, functions or organizations. In contrast, commercial off-the-shelf software (COTS) is designed for a broad set of requirements, allowing it to be packaged and commercially marketed and distributed.

Developing software typically involves the following steps:

- **Selecting a methodology** to establish a framework in which the steps of software development are applied. It describes an overall work process or roadmap for the project. Methodologies can include Agile development, DevOps, Rapid Application Development (RAD), Scaled Agile Framework (SAFe), Waterfall and others.

- **Gathering requirements** to understand and document what is required by users and other stakeholders.

- **Choosing or building an architecture** as the underlying structure within which the software will operate.

- **Developing a design** around solutions to the problems presented by requirements, often involving process models and storyboards.

- **Building a model** with a modeling tool that uses a modeling language like SysML or UML to conduct early validation, prototyping and simulation of the design.

- **Constructing code** in the appropriate programming language. This involves peer and team review to eliminate problems early and produce quality software faster.

- **Testing** with pre-planned scenarios as part of software design and coding – and conducting performance testing to simulate load testing on the application.

- **Managing configuration and defects** to understand all the software artifacts (requirements, design, code, tests) and build distinct versions of the software. Establish quality assurance priorities and release criteria to address and track defects.

- **Deploying** the software for use and responding to and resolving user problems.

- **Migrating data** to the new or updated software from existing applications or data sources if necessary.

- **Managing and measuring the project** to maintain quality and delivery over the application lifecycle, and to evaluate the development process with models such as the Capability Maturity Model (CMM).

The steps of the software development process fit into application lifecycle management (ALM). The IBM engineering management solution is a superset of ALM that enables the management of parallel mechanical, electrical and software development, and involves:

- Requirements analysis and specification;
- Design and development;
- Testing;
- Deployment;
- Maintenance and support.

Software development process steps can be grouped into the phases of the lifecycle, but the importance of the lifecycle is that it recycles to enable continuous improvement. For example, user issues that surface in the maintenance and support phase can become requirements at the beginning of the next cycle.

1.3 Development Tools

It is no longer surprising to hear that Python is one of the most popular languages among developers and in the data science community. While there are numerous reasons behind Python's popularity, it is primarily because of two core reasons:

- Python has a very simple syntax – almost equivalent to mathematical syntax – and hence it can be easily understood and learned.
- Python offers extensive coverage (libraries, tools, etc.) for scientific computing and data science.

There are numerous reasons to use Python for data science. For now we'll discuss some of the Python tools most widely used by developers, coders and data scientists across the world. These tools are useful for many different purposes if you know how to use them correctly. So, without further delay, let's look at the best Python tools out there!

Scikit-Learn: This is an open-source tool designed for data science and machine learning. It is extensively used by developers, machine learning

(ML) engineers and data scientists for data mining and data analysis. One of the greatest features of Scikit-Learn is its remarkable speed in performing different benchmarks on toy datasets. The primary characteristics of this tool are classification, regression, clustering, dimensionality reduction, model selection and preprocessing. It offers a consistent and user-friendly application program interface (API) along with grid and random searches.

Keras: This is an open-source, high-level, neural network library written in Python. It is highly suited for ML and deep learning. Keras is based on four core principles – user-friendliness, modularity, easy extensibility and working with Python. It allows you to express neural networks in the easiest way possible. Since Keras is written in Python, it can run on top of popular neural network frameworks like TensorFlow, CNTK and Theano.

Theano: This is a Python library designed explicitly for expressing multi-dimensional arrays. It allows you to define, optimize and evaluate mathematical computations comprising multi-dimensional arrays. Some of its most unique features include its tight integration with NumPy, transparent use of a graphics programming unit (GPU), efficient symbolic differentiation, speed and stability optimization, dynamic C code generation and extensive unit-testing, to name a few.

SciPy: This is an open-source Python-based library ecosystem used for scientific and technical computing. It is extensively used in the fields of mathematics, science and engineering. SciPy leverages other Python packages, including NumPy, IPython and Pandas, to create libraries for common math and science-oriented programming tasks. It is an excellent tool for manipulating numbers on a computer and generating visualized results.

1.3.1 Advanced Python Tools

Selenium: This is undoubtedly one of the best Python development tools. It is an open-source automation framework for web applications. With Selenium, you can write test scripts in many other programming languages, including Java, C#, PHP, Perl, Ruby and .NET.

Furthermore, you can perform tests from any browser (Chrome, Firefox, Safari, Opera and Internet Explorer) in all of the three major operating systems – Windows, macOS and Linux. You can also integrate Selenium with tools like JUnit and TestNG for managing test cases and generating reports.

Robot Framework: This is another open-source generic test automation framework designed for acceptance testing and acceptance test-driven development (ATTD). It uses tabular test data syntax and is keyword-driven. Robot Framework integrates many frameworks for different test automation requirements.

You can expand the framework's abilities by further integrating it with Python or Java libraries. Robot Framework can be used not only for web app testing but also for Android and iOS test automation.

TestComplete: This is software that supports web, mobile and desktop automation testing. However, you must acquire a commercial license to be able to use it. TestComplete also allows you to perform keyword-driven testing, just like Robot Framework. It comes with an easy-to-use record and playback feature.

It supports many scripting languages, including Python, VBScript and C++ script. Just like Robot Framework, software testers can perform keyword-driven testing. A noteworthy feature of this Python tool is that its GUI object recognition abilities can both detect and update UI objects. This helps reduce the efforts required to maintain test scripts.

1.3.2 Web Scraping Python Tools

Beautiful Soup: This is a Python library for extracting data from HTML and XML files. You can integrate it with your preferred parser to leverage various Pythonic idioms for navigating, searching and modifying a parse tree. The tool can automatically convert incoming documents to Unicode and outgoing documents to UTF-8 and is used for projects like screen-scraping. It is a great tool that can save you hours of work [2].

LXML: This is a Python-based tool designed for C libraries – libxml2 and libxslt. It is highly feature-rich and one of the most easy-to-use libraries for processing XML and HTML in Python. It facilitates safe and convenient access to libxml2 and libxslt libraries by using the ElementTree API.

Scrapy: This is an open-source and collaborative framework written in Python. Essentially, it is an application framework used for developing web spiders (the classes that a user defines) that crawl web sites and extract data from them.

Scrapy is a fast, high-level web crawling and scraping framework that can also be used for many other tasks like data mining and automated testing. It can efficiently run on all three major operating systems, that is, Windows, macOS and Linux.

Urllib: This is a Python package that is designed for collecting and opening URLs. It has various modules and functions to work with URLs. For instance, it uses: "urllib.request" for opening and reading URLs that are mostly HTTP; "urllib.error" to define the classes for exceptions raised by urllib.request; "urllib.parse" to define a standard interface to fragment URL strings in components; and a "urllib.robotparser" function to create a single class. The higher-ranking language version is the source code. The translated machine language code is called the goal program. The converter converts the source code to the targeted computer's script.

Higher-level languages are beautiful: the same source code from Python will operate on various target platforms. A Python interpreter should be given for the target framework but for all major programming systems various interpreters are necessary. Therefore, it is not in a specific engine language that the human programmer thinks about writing the solution to a question in Python.

Programmers have access to a range of tools to improve the development of software. Some traditional instruments include:

- **Publishers**. A programmer can access the source code of the software and copy it to files in an editor. Some software editors improve the efficiency of programmers with colors to highlight language capabilities. Language syntax refers to how well-formed sentences are arranged in parts of the language. "The large child moves quickly to the entrance" uses correct English grammar. "Child, the large is moving to the entrance quickly" is not syntactically correct. It uses the same vocabulary as the initial sentence but does not obey the rules of English structure. In the same way, programming languages are subject to strict syntax rules to create well-formed programs. Some syntax editors use colors or other special notes to warn of syntax mistakes by programmers before compiling the software.

- **Python compilers**. The source code is translated into the target code by a compiler. The goal code can also be a programming language or an embedded program for a single device. Another source language could be the target code; C++, another higher-level language, is interpreted by, for example, the earliest C++ compiler.

1.4 learning about the Python Compiler

The C code has already been translated to an executable application by a C compiler. (C++ compilers now directly translate C++ into the language of the machine.) But Sublime Text is a popular code editor that supports many languages including Python. It's fast, highly customizable and has a huge community. It provides basic built-in support for Python when you install it. However, you can install packages such as debugging, auto-completion and code linting:

- **Interpreters**. An interpreter is like a translator in which the source is interpreted into the computer language at a higher level. However, it behaves differently. Whereas a compiler creates an interactive application that will operate regularly without any further translation, a compiler converts source code statements to a computer language when the software is operating. There's no need for a compiled program to operate, so any time it's performed, an encoded program will be processed. Typically compiled programs are faster than translated programs, as translation only takes place once. On the other hand, translated applications will be executed on a computer with a compatible interpreter; they don't have to be recompiled to operate on another device. For example, Python is mainly used as an interpreted language, but compilers are available for it. Interpreted languages are more suited for complex and exploratory development, which is perfect for beginner programmers.

- **Debuggers**. A debugger enables programmers to run a program simultaneously and to verify the source code line currently running if the values of the variable and other system elements shift as planned. Debuggers are important for the location of bugs and the repair of error-containing programs.

- **Profilers**. To assess the output of a system, a profiler is used. This indicates how much a portion of the system is run during a given execution and how long it takes for that component to operate. Profilers can also be used to test and ensure that all technology in a system, as checked, is properly used. This is called reporting. Software typically does not function until it has been launched as consumers access a component of the software that has not been implemented during development. Profiling is primarily intended to identify the elements of a program that can be improved to make it run quicker.

Included developing environments (IDEs) are used by many developers. For a robust system, an IDE contains templates, debuggers and other programming aids. For example, commercial IDEs include the 2010 Visual Studio of Microsoft, the 2010 Eclipse IDE of the Eclipse Foundation and the XCode of Apple. For Python, IDLE is a simple IDE. The programming process in all but trivial applications is not automated given the many technologies (and the statements of tool vendors). Good instruments are useful and certainly increase developers' productivity, but software cannot be written. Sound critical reasoning, creativeness, common sense and programming knowledge have not been replaced.

1.5 Python History

Guido van Rossum created the Python programming language at the end of the 1980s. Unlike other modern languages, such as C, C++, Java and C#, Python is aimed at constructing a simple but powerful syntax. The language was implemented in December 1989 in the Netherlands as a successor to ABC and was capable of exception handling and interfacing with the

FIGURE 1.1
Python scientist Guido van Rossum.

Amoeba operating system (see Figure 1.1). Van Rossum is Python's principal author, and his continuing central role in deciding its direction is reflected in the title given to him by the Python community: Benevolent Dictator for Life. (However, van Rossum stepped down as leader on July 12, 2018.) The name Python came from the BBC TV show *Monty Python's Flying Circus*.

Python 2.0 was released on October 16, 2000, with many major new features, including a cycle-detecting garbage collector (in addition to reference counting) for memory management and support for Unicode. However, the most important change was to the development process itself, with a shift to a more transparent and community-backed process.

Python 3.0, a major, backwards-incompatible release, was released on December 3, 2008 after a long period of testing. Many of its major features were also backported to the backwards-compatible, though now-unsupported, Python 2.6 and 2.7.

In the app creation in Facebook, Twitter, CERN, Industrial Light & Magic and NASA, Python is included. Experienced programmers may do amazing stuff with Python, but the strength of it is to encourage them to make innovations and to work with fascinating problems more quickly than with other languages that are more complex and have a steeper knowledge slope. Improved Python features include links to Microsoft Windows' latest updates.

1.6 Python Installation

1.6.1 Step 1: Select Version of Python to Install

The installation procedure involves downloading the official Python .exe installer and running it on your system.

The version you need depends on what you want to do in Python. For example, if you are working on a project coded in Python version 2.6, you probably need that version. If you are starting a project from scratch, you have the freedom to choose.

If you are learning to code in Python, we recommend you *download the latest versions of Python 2 and 3*. Working with Python 2 enables you to work on older projects or test new ones for backward compatibility.

1.6.2 Step 2: Download Python Executable Installer

- Open your web browser and navigate to the Downloads for Windows section of the official Python website.

- Search for your desired version of Python. At present, the latest Python 3 release is version 3.10.4, while the latest 3 release is version 3.9.3.

- Select a link to download either the **Windows x86-64 executable installer** or the **Windows x86 executable installer**. The download is approximately 25 MB (see Figure 1.2).

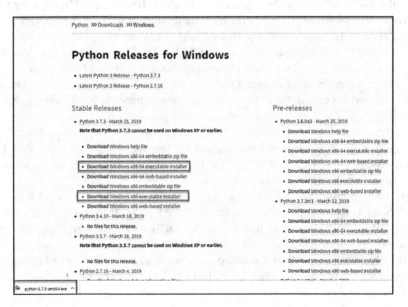

FIGURE 1.2
Python Release Versions.

1.6.3 Step 3: Run Executable Installer

- Run the **Python Installer** once downloaded. (In this example, we have downloaded Python 3.7.3.)
- Make sure you select the **Install launcher for all users** and **Add Python 3.7 to PATH** checkboxes. The latter places the interpreter in the execution path. For older versions of Python that do not support the **Add Python to Path** checkbox, see Step 6.
- Select **Install Now** – the recommended installation options are shown in Figure 1.3.

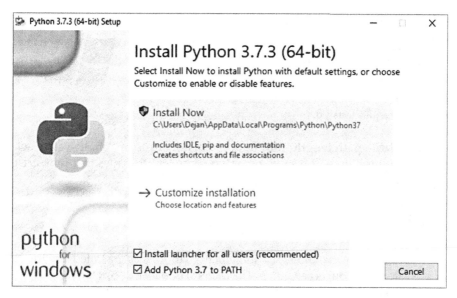

FIGURE 1.3
Python installation process.

For all recent versions of Python, the recommended installation options include **Pip** and **IDLE**. Older versions might not include such additional features.

- The next dialog box will prompt you to select **Disable path length limit**. Choosing this option will allow Python to bypass the 260-character MAX_PATH limit. Effectively, it will enable Python to use long path names (see Figure 1.4).

FIGURE 1.4
Python successfully installed.

The **Disable path length limit** option will not affect any other system settings. Turning it on will resolve potential name length issues that may arise with Python projects developed in Linux.

1.6.4 Step 4: Verify Python Was Installed on Windows

- Navigate to the directory in which Python was installed on the system. In our case, it is **C:\Users**_Username_**\AppData\Local\ Programs\Python\Python37** since we have installed the latest version.
- Double-click **python.exe**.
- The output should be similar to what you can see in Figure 1.5.

1.6.5 Step 5: Verify Pip Was Installed

If you opted to install an older version of Python, it is possible that it did not come with Pip preinstalled. Pip is a powerful management system for Python software packages. So, make sure that you have it installed [3].

We recommend using Pip for most Python packages, especially when working in virtual environments. To verify whether Pip was installed:

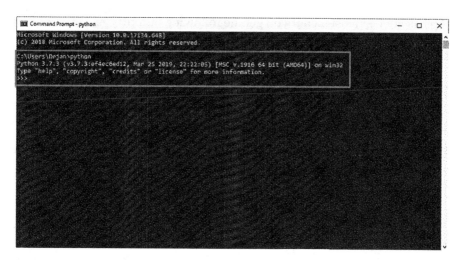

FIGURE 1.5
Python path on DOS.

- Open the **Start** menu and type "**cmd.**"
- Select the **Command Prompt** application.
- Enter **pip -V** in the console. If Pip was installed successfully, you should see the display in Figure 1.6.

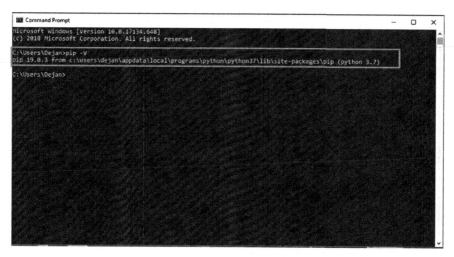

FIGURE 1.6
Python check version.

Pip has not been installed if you get the following output:

'pip' is not recognized as an internal or external command, Operable program or batch file.

If your version of Python is missing Pip, see How to Install Pip to Manage Python Packages on Windows.

1.6.6 Step 6: Add Python Path to Environment Variables (Optional)

We recommend you go through this step if your version of the Python installer does not include the **Add Python to PATH** checkbox or if you have not selected that option.

Setting up the Python path to system variables alleviates the need for using full paths. It instructs Windows to look through all the PATH folders for "python" and find the install folder that contains the python.exe file (see Figure 1.7).

- Open the **Start** menu and start the **Run** app.

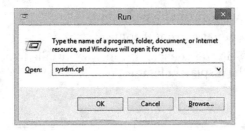

FIGURE 1.7
System property window.

- Type **sysdm.cpl** and click **OK**. This opens the **System Properties** window.
- Navigate to the **Advanced** tab and select **Environment Variables**.
- Under **System Variables**, find and select the **Path** variable.
- Click **Edit**.
- Select the **Variable value** field. Add the path to the **python.exe** file preceded with a **semicolon (;)**. For example, in Figure 1.8, we have added **C:\Python34**.
- Click **OK** and close all windows.

By setting this up, you can execute Python scripts like this: Python script.py
Instead of this: C:/Python34/Python script.py
As you can see, it is cleaner and more manageable.

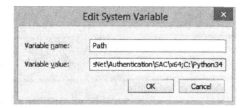

FIGURE 1.8
Python set path.

1.6.7 Step 7: Install virtualnv (Optional)

You have Python, and you have Pip to manage packages. Now, you need one last software package: **virtualnv**. Virtualnv enables you to create isolated local virtual environments for your Python projects.

Python software packages are installed system-wide by default. Consequently, whenever a single project-specific package is changed, it changes for all your Python projects. You need to avoid this; having separate virtual environments for each project is the easiest solution.

To install virtualnv:

- Open the **Start** menu and type "**cmd.**"
- Select the **Command Prompt** application.
- Type the following pip command in the console:

```
C:\Users\Username> pip install virtualenv
```

Upon completion, virtualnv is installed on your system.

1.7 How to Write a Python Program

Figure 1.9 demonstrates how to navigate the Microsoft Window IDLE in the Inbox line.

We have discussed the steps to install the latest release of Python. Installing Python also installs its IDLE, which is an IDE (an interactive integrated development environment) used to code in Python, which highlights all the keywords and has a Python Shell, also known as an interpreter, is responsible for the execution of the Python programs, displays their output and notifies us about the errors in a program.

In order to execute your first python program using the IDLE IDE, we need to go to the folder where it has been installed and execute it (see Figure 1.10).

FIGURE 1.9
Open IDLE from start menu button in windows operating system.

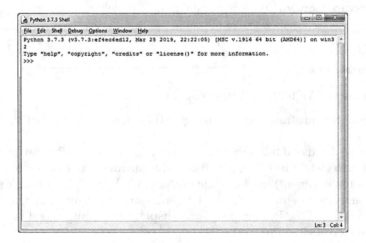

FIGURE 1.10
Python environment.

Double-clicking the file **idle.bat** in this folder, executes the IDLE IDE and opens the window of Python Shell with the prompt >>> (where you can type in a code to be executed).

There are two ways to code a Python program using IDLE:

- By coding directly into the **Python Shell** of **IDLE**;
- By coding in a separate file.

In both cases, the Python Shell (interpreter) will display the output or the errors associated with the program. Let us see both situations with an example by printing the customary message "Hello World!"

In Python Shell, we could type in some Python code, which will be executed in the sequence we type it in; its output is displayed. Let us type in a small program at the prompt of Python Shell (i.e. >>>), which prints the string message "Hello World", as shown in Figure 1.11.

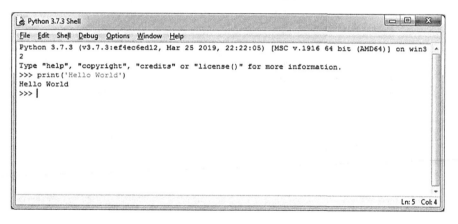

FIGURE 1.11
Python coding environment.

Using the Python Shell window, we can create a *new separate file* by clicking on **File-> New File** menu options are in the top menu with code in it. Let us name this file **first.py**.

Note: All console programs in Python are stored with a **.py** extension, whereas all GUI Python programs are stored with a **.pyw** extension.

By default, the programs coded in a separate file using IDLE will be stored in the same directory where IDLE is stored:

```
C:\Users\Your_UserName\AppData\Local\Programs\Python\Python37\
Lib\idlelib
```

This editor allows programs to be saved and later updated conveniently. The editor knows Python's syntax and uses multiple colors to illustrate the specific components of a system. A lot of the production activity takes place in the printer (see Figure 1.12).

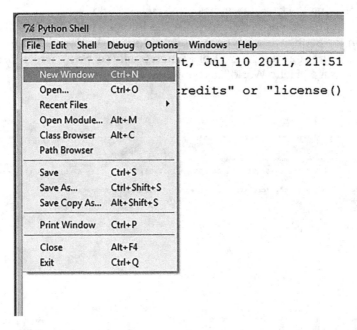

FIGURE 1.12
Python's open new code window.

After writing a program you can store it in a directory of your choice; and that's what we are going to do now. We will save the file of our first Python program in the location **D:/Python** (as shown in Figure 1.13), and we are going to name it **first.py**.

To save the program script, scroll down the file options and choose the save option (see Figure 1.14).

To execute this code in a separate file, we have three options:

- We could click on the menu options on top of this window: **Run -> Run Module**;

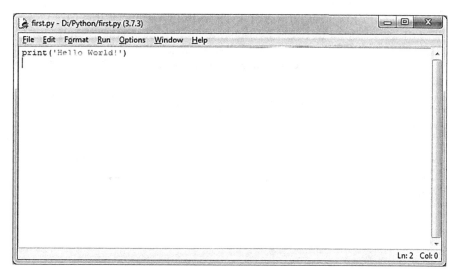

FIGURE 1.13
Run scripting environment.

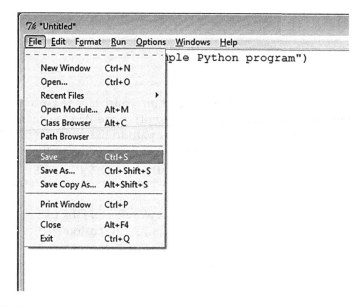

FIGURE 1.14
Save an IDLE editor generated file.

- We could just press the **ALT + F5** in the window of this program which is coded in IDLE. Doing so will open the **Python Shell** window, displaying the output of the code which was stored in this separate file or the errors associated with it, as shown in Figure 1.15.

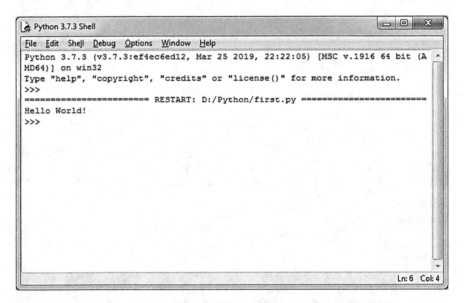

FIGURE 1.15
Output window.

- We could also execute a Python program (which is coded in a separate file) using the **CMD** (command prompt) of the Windows operating system, by just changing the current path to the directory in which we have stored our Python program and by typing in the name of the program with its **.py** extension. This launches the interpreter and executes the program in the console window, as shown in Figure 1.16.

Since the interactive shell does not have a way to store the entered file, it is not the safest way to compose broad programs. The customizable IDLE shell is good for short Python code fragments.

The Figure 1.13 (simpleprogram.py) collection includes just one code line:

```
>>>print("Hello World!")
```

This is a comment by Python. A statement is an order performed by the translator. This declaration prints the Python software message on the computer. The key execution unit in the program Python is a statement. Declarations can be separated into large blocks called "block", and block statements may

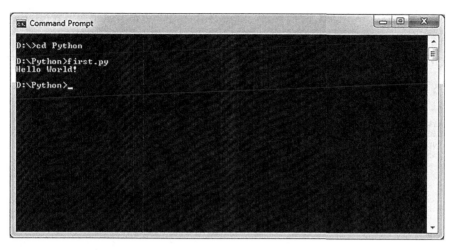

FIGURE 1.16
Python run on DOS prompt.

be more complicated. Blocks consist of higher frameworks such as tasks and processes [4]. The statement

```
>>>print("Hello World!")
```

utilizes a built-in print function. Python has numerous types of these statements which are used to construct systems, as we will see in the subsequent pages.

Program 1.1

Further complex services inherit several statements. Six typed sentences display a file on the computer in description python program (arrow 1.2 py):

```
print ("    *    ")
print ("    *    ")
print ("    *    ")
print (" ***** ")
print ("   ***   ")
print ("    *    ")
```

Every print statement in the Python program (arrow 1.2. py) "draws" a horizontal arrow. The picture of the arrow comes from all the horizontal slices piled on top of each other. The statements are a document of Python code. Particularly before the beginning of each sentence no whitespace (spots or tabs) will be produced. In Python, comments must be indented and rendered correctly.

Output

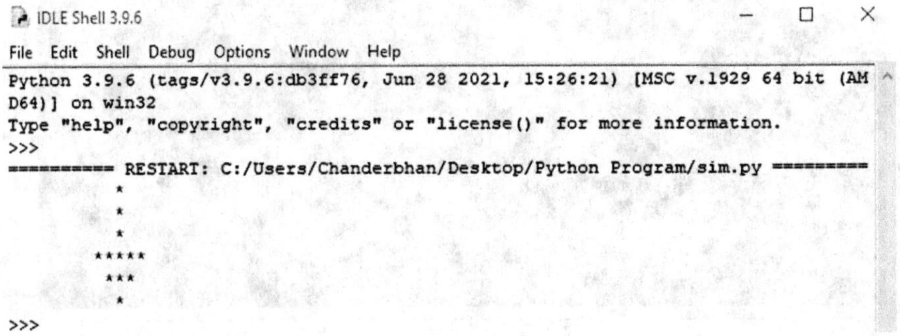

```
IDLE Shell 3.9.6                                                    —    □    ×
File  Edit  Shell  Debug  Options  Window  Help
Python 3.9.6 (tags/v3.9.6:db3ff76, Jun 28 2021, 15:26:21) [MSC v.1929 64 bit (AM ^
D64)] on win32
Type "help", "copyright", "credits" or "license()" for more information.
>>>
========== RESTART: C:/Users/Chanderbhan/Desktop/Python Program/sim.py ==========
        *
        *
        *
      *****
      ***
        *
>>>
```

FIGURE 1.17
Output of arrow 1.2.py.

When you seek to reach the dynamic shell of each line, one at a time, the output of the software is combined with your comments. The safest approach in this situation is to type the software into an editor, saving the application you write in a register, and running it. We use an editor for much of the time to access and execute Python. The dynamic interpreter is most useful for testing short Python-code snippets [5].

If you try to insert a space in the interactive shell before a sentence, you get the output shown in Figure 1.18.

Statement

```
>>> Print('hello python')
```

Output

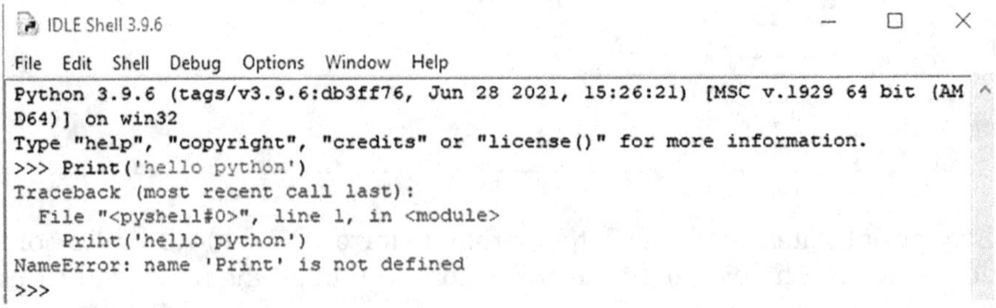

```
IDLE Shell 3.9.6                                                    —    □    ×
File  Edit  Shell  Debug  Options  Window  Help
Python 3.9.6 (tags/v3.9.6:db3ff76, Jun 28 2021, 15:26:21) [MSC v.1929 64 bit (AM ^
D64)] on win32
Type "help", "copyright", "credits" or "license()" for more information.
>>> Print('hello python')
Traceback (most recent call last):
  File "<pyshell#0>", line 1, in <module>
    Print('hello python')
NameError: name 'Print' is not defined
>>>
```

FIGURE 1.18
Output of simple print statement.

When the application includes the extraneous indentation, the interpreter makes a related error as we seek to execute the saved Python file.

1.8 Conclusion

In this chapter we have discussed Python in general, and installed it in a system. We have installed a text editor to make it easier to write Python code and run snippets of it in a terminal session, and we ran a first actual program, hello_world.py. The Python programming language is a suitable choice for learning and real world programming. We have discussed the characteristics, features and types of programming support offered by Python. According to these characteristics, we have found Python to be a fast, powerful, portable, simple and open source language that supports other technologies. We have also discussed the latest applications of Python, some of its popular concepts and its origin.

2

Data Types and Variables

2.1 Python Integer Values

In Python, integers are zero, positive or negative whole numbers without a fractional part and with unlimited precision, for example 0, 100, -10. The following are valid integer literals in Python:

```
>>> 0
0
>>> 100
100
>>> -10
10
>>> 1234567890
1234567890
>>> y=500000000000000000000000000000000000000000000000000000000
0000
500000000000000000000000000000000000000000000000000000000
```

Integers can be binary, octal and hexadecimal values:

```
>>> 0b11011000 # binary
216
>>> 0o12 # octal
10
>>> 0x12 # hexadecimal
15
```

All integer literals or variables are objects of the int class. Use the type ()
method to get the class name:

```
>>>type(100)
<class 'int'> # type of x is int

>>> x=1234567890
>>> type(x)
<class 'int'> # type of x is int

>>> y=5000000000000000000000000000000000000000000000000000000000
0000
>>> type(y) # type of y is int
<class 'int'>
```

Leading zeros in non-zero integers are not allowed, for example 000123 is an
invalid number, 0000 is 0:

```
>>> x=01234567890
SyntaxError: invalid token
```

Python does not allow a comma as a number delimiter. Use underscore "_"
as a delimiter instead:

```
>>> x=1_234_567_890
>>> x
1234567890
```

Note that integers must be without a fractional part (decimal point). If it
includes a fractional then it becomes a float:

```
>>> x=5
>>> type(x)
<class 'int'>
>>> x=5.0
>>> type(x)
<class 'float'>
```

The int () function converts a string or float to int:

```
>>> int('100')
100
>>> int('-10')
-10
>>> int('5.5')
5
>>> int('100', 2)
4
```

- **Binary**

  A number having **0b** with eight digits in combinations of 0 and 1 represent the binary numbers in Python. For example, 0b11011000 is a binary number equivalent to integer 216:

```
>>> x=0b11011000
>>> x
216
>>> x=0b_1101_1000
>>> x
216
>>> type(x)
<class 'int'>
```

- **Octal**

  A number having **0o** or **0O** as a prefix represents an octal number. For example, 0O12 is equivalent to integer 10:

```
>>> x=0o12
>>> x
10
>>> type(x)
<class 'int'>
```

- **Hexadecimal**

  A number with **0x** or **0x** as a prefix represents a hexadecimal number. For example, 0x12 is equivalent to integer 18:

```
>>> x=0x12
>>> x
18
>>> type(x)
<class 'int'>
```

- **Float**

  In Python, floating point numbers (floats) are positive and negative real numbers with a fractional part denoted by the decimal symbol . or the scientific notation E or e, for example 1234.56, 3.142, −1.55, 0.23:

```
>>> f=1.2
>>> f
1.2
>>> type(f)
<class 'float'>
```

Floats can be separated by the underscore _, for example 123_42.222_013 is a valid float:

```
>>> f=123_42.222_013
>>> f
12342.222013
```

Floats has a maximum size depending on your system. A float beyond its maximum size is referred to as "inf", "Inf", "INFINITY", or "infinity". Float 2e400 will be considered as infinity for most systems:

```
>>> f=2e400
>>> f
inf
```

Scientific notation is used as a short representation to express floats having many digits. For example, 345.56789 is represented as 3.4556789e2 or 3.4556789E2:

```
>>> f=1e3
>>> f
1000.0
>>> f=1e5
>>> f
100000.0
>>> f=3.4556789e2
>>> f
345.56789
```

Use the float() function to convert string, int to float:

```
>>> float('5.5')
5.5
>>> float('5')
5.0
>>> float(' -5')
-5.0
>>> float('1e3')
1000.0
>>> float('-Infinity')
-inf
>>> float('inf')
inf
```

2.1.1 Complex Numbers

A complex number is a number with real and imaginary components. For example, 5 + 6j is a complex number where 5 is the real component and 6 multiplied by j is an imaginary component:

```
>>> a=5+2j
>>> a
(5+2j)
>>> type(a)
<class 'complex'>
```

You must use j or J for the imaginary component. Using another character will produce a syntax error:

```
>>> a=5+2k
SyntaxError: invalid syntax
>>> a=5+j
SyntaxError: invalid syntax
>>> a=5j+2j
SyntaxError: invalid syntax
```

To make the dynamic shell function as a convenient addition tool, Python uses the + symbol for integers:

```
>>> 8+8
16
>>> 4+5+2+3+8+9
31
>>>print (4+5+2+3+8+9)
31
```

The last line tested illustrates how the + sign may be used to add values in a typed sentence to a Python application [6].
   Remember what occurs when we use add and quotes in an integer:

```
>>>print(4+5+2+3+8+9)
31
>>>"225"
'225'
>>>'225'
'225'
```

Note how special the interpreter's output is. The term '225' is an example of a string value. A character set is a sequence. A full character is typically found in strings:

```
>>>"Blue"
'Blue'
>>>'Blue'
'Blue
```

Python understands the correct form of restricting a string length for both single (") and double quotes (""). When the start of a string value is indicated by a single quotation, the end of the string must be specified. Likewise, if used, double quotes will be used in pairs.

```
>>> 'PQRS'
' PQRS'
>>> " PQRS "
' PQRS'
>>> ' PQRS "
    File "<stdin>", line 1 ' PQRS "
  ^SyntaxError: EOL while scanning string literal
>>> " PQRS'
    File "<stdin>", line 1 " PQRS'
^SyntaxError: EOL while scanning string literal
```

Performance of the interpreter uses separate quotations mostly, but allows single quotes or duplicates as legitimate inputs [7].

Find the following series of interactions:

```
>>> 56
56
>>> "56"
'56'
>>> ' 56'
' 56'
>>> "Hello"
' Hello'
>>> ' Hello'
' Hello'
>>> Hello
Traceback (most recent call last):
File "<stdin>", line 1, in <module>
NameError: name ' Hello' is not defined
```

The interpreter did not recognize Hello (without quotes) because the quotation marks were missing. The terms 56 and '56' are special, it is important to remember. That is a complete sequence of statements relate to a set of statements. Both terms include the Python form. A word style indicates the form of speech. Often the form of an expression is referred to as a statement. At this stage, only integers and sequences have been used as variables.

The built-in form feature demonstrates the nature of every expression in Python [8]:

```
>>> type (869)
<class 'int'>
>>> type ('869')
<class 'str'>
```

Python compares the word int to integers and str to sequences of strings.

The incorporated int function transforms the entire variable representation into a real integer and transforms an integral expression:

```
>>> 81
81
>>> str(81)
' 81'
>>> '54'
'54'
>>> int('54')
54
```

The word str (81) calculates the value of the string '81' while int ('54') calculates the integer number 54. The int method added to an integer essentially calculates the numerical type; str is used for a variable in the same way as the initial array:

```
>>> int(45)
45
>>> str('Robin')
' Robin'
```

As one might guess, a programmer does not need to try these changes. Also, str('45') is harder to say clearly than 45, and therefore it cannot be shown or proven that the value of the str and int functions before the numbers that change/things that change have been applied in Section 10.2. Any integer looks like a number or a digit but any digit or number written in double and single quotes is different:

```
    >>> str (65536)
'65536'
>>> int ('Hi')
Traceback (most recent call last): .
    File "<pyshell#5>", line 1, in <module>
        int ('Hi')
ValueError: invalid literal for int () with base 10: 'Hi'
>>> int ('52.23')
Traceback (most recent call last):
    File "<pyshell#6>", line 1, in <module>
        int ('52.23')
ValueError: invalid literal for int () with base 10:
'52.23'
```

Alas, 65536 are not true entered expressions in Python. Simply stated, if the contents of the string are like a real integer number (those features that render it so), you can easily use the int method to generate the specified integer.

For strings, the plus operator (+) functions differently:

```
>>> 892 + 800
1692
>>> '892' + '800'
'892800'
>>> 'pqrs' + 'abcd'
'pqrsabcd'
```

The outcome is very distinct. As you can see '892' + '800' and 'pqrs' + 'abcd' in the phrase 892 + 800 shows the addition result. In a method defined as a concatenation, the plus operator separates two numbers. It is not allowed to combine all styles directly:

```
>>> '892'+800
Traceback (most recent call last):
    File "<pyshell#8>", line 1, in <module>
        '892'+800
TypeError: can only concatenate str (not "int") to str
>>> 892+'800'
```

```
Traceback (most recent call last):
    File "<pyshell#9>", line 1, in <module>
        892+'800'
TypeError: unsupported operand type(s) for +: 'int' and
'str'
```

In '892' + 800, one number is a string and the other number is an int; the combination of the two items does not fulfill the output, which means it will show an error.

But the functions int and str will help:

```
>>> 892 + int('800')
1692
>>> '892' + str(800)
'892800'
```

In 892 + int('800'), the first number is in integer format, but the other is written as a string in inverted commas, using pre-data type to convert strings accordingly. When str is used in the pre-initialization of variables its behavior changes.

The style function will decide which expressions are the most complex and show type def:

```
>>> type(82)
<class 'int'>
>>> type ('82')
<class 'str'>
>>> type (52+87)
<class 'int'>
>>> type ('52'+'87')
<class 'str'>
>>> type (int('8')+int('9'))
<class 'int'>
>>> 2000+400+68
2468
```

In numerical values in Python, commas cannot exist. The sum of 2,000, 400 and 68 will be 2468, rather than 2,468.

In arithmetic, the integer is unlimited; in other words, there are infinite logical integers. For random numbers in Python, the larger the integer, the more resources it requires. In reality, Python entries are constrained by how much memory is accessible, though, as a machine, this is finite (and the operating system also restricts memory by running software).

2.2 Variables and Assignment

Numbers or things that change are variables in math, where letters stand for numbers. The same applies to Python, except that things other than numbers can also be described. Program 2.1 (simplevariable.py) needs an integer value index and then outputs the changeable value [9].

```
>>>x = 90
>>>print(x)
90
```

There are two statements in Program 2.1 (simplevariable.py).

A statement of initialization of the variable: $x = 90$. This is a declaration of assignment of a numerical value. The symbol = is known as the assignment operator. The argument refers to the variable x with the integer value 90. In other words, this connects the x to the value of 90. As often as appropriate, an attribute may be allocated or reassigned. If an element of another form is reallocated, the type of a variable may change.

This declaration demonstrates the present state of the variable x. Recall the lack of quote marks, which is very significant. When x is 90, the sentence

```
print (x)
```

prints 90, variable x=10 means 10 is assigned to x as a
declaration

```
print ('x')
```

prints x, an atomic letter x post.

The assignment operator (=) is different from that of logical equality. In algebra, "=" affirms that the left-hand statement is equivalent to the right-hand statement. In Python, "=" assigns the value of the statement to the right of the element to that of the left. It is easier to say $x = 12$ than "x is 12". In arithmetic it is symmetrical: if $x = 12$, we know $12 = x$. There is no such symmetry in the Python statement

```
12 = x
```

Trying to reassign actual integer value 5, but 5 is still 5 and cannot be modified. Such a comment is bound to produce an error.

```
>>> x=12
>>> x
12
>>> 12=x
SyntaxError: cannot assign to literal
```

As Program 2.2 (multipleassign.Py) indicates, variables can be reassigned with various values as appropriate.

```
>>> x = 70
>>> print ('x = ' + str (x))
x = 70
>>> x = 80
>>> print ('x = ' + str (x))
x = 80
>>> x = 50
>>> print ('x = ' + str (x))
x = 50
```

Notice that in Program 2.2 (multipleassign.py), the print statements are similar, but if the system is executed, the statements show separate results (as a method, and not in an interactive shell):

$$
\begin{aligned}
x &= 70 \\
x &= 80 \\
x &= 50
\end{aligned}
$$

The variable x is int-like because an integer value is assign to it. Recall how statement x uses `str` to manage it as a Python program (multipleassign.py) such that variable concatenation is done by a + operator:

```
print ('y = ' + str(y))
```

Since the plus operator cannot be shown as above, it is used for a combination of strings and integer operands, hence the term `'y = '+ y` is not valid.

Program 2.3 (multipleassign2.py) includes a variant of the program that generates the same output.

```
>>>x = 200
>>>print ('x -', x)
x = 200
>>>x = 74
>>>print ('x =', x)
x = 74
>>>x = 52
>>>print ('x =', x)
x = 52
```

The print comment in this version

```
print ('x =', x)
```

illustrates the print function accepting two parameters. The first parameter is the string 'x =', and the second parameter is the variable x bound to an integer value. The print function allows programmers to pass multiple expressions to print, each separated by commas. The elements within the parentheses of the print function comprise what is known as a comma-separated list. The print function prints each element in the comma-separated list of parameters. The print function automatically prints a space between each element in the list so they do not run together.

In a single declaration, a programmer may assign multiple variables. Program 2.4 (valueassign.py) shows how:

```
x,y,z = 100, -45, 0
print ('x=',x,'y=',y,'z=',z)
```

Program 2.3 generates output (valueassign.py):

x = 100	y= -45	z = 0

A tuple is a set of words separated by commas. In the statement of assignment

```
x, y, z = 100, -45, 0
```

x, y, z is one tuple, and 100, −45, 0 is another tuple. Tuple assignment works as follows. The first variable in the tuple on the left side of the assignment operator is assigned the value of the first expression in the tuple on the right side (effectively x = 100). Similarly, the second variable in the tuple on the left side of the assignment operator is assigned the value of the second expression in the tuple on the right side (in effect y = −45). z gets the value 0.

An assignment statement binds a variable name to an object. We can visualize this process with a box and arrow as shown in Figure 2.1.

We name the box with the variable's name. The arrow projecting from the box points to the object to which the variable is bound. Figure 2.2 shows how variable bindings change as the following sequence of Python is executed:

Importantly, the statement a = b means that a and b both are bound to the same numeric object. Note that reassigning b does not affect a's value.

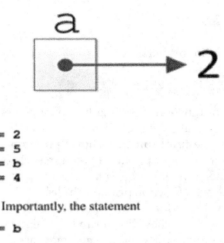

FIGURE 2.1
Binding a variable to an object.

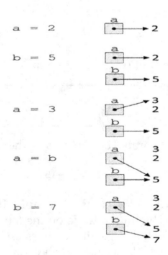

FIGURE 2.2
Variable bindings stage how it changes as program runs.

Not only may a variable's value change during its use within an executing program, the type of a variable can change as well:

```
a=10
print('First, variable a has value',a,'and type',type(a))
b='abc'
print('Now, variable b has value',b,'and type',type(b))
```

The result is:

```
First, variable a has value 10 and type <class 'int'>
Now, variable a has value ABC and type <class 'str'>
```

Most variables maintain their original type throughout a program's execution. A variable should have a specific meaning within a program, and its meaning should not change during the program's execution. While not always the case, sometimes when a variable's type changes its meaning changes as well.

2.3 Identifiers in Python

Good variable names make programs more readable by humans. Since programs often contain many variables, well-chosen variable names can render an otherwise obscure collection of symbols more understandable.

Python has strict rules for variable names. A variable name is one example of an identifier. An identifier is a term used to name objects and is a built-in function or reserved keyword as shown in Table 2.1 [10].

TABLE 2.1

Python Built-in Function

and	del	from	None	try
as	elif	global	nonlocal	True
assert	else	if	not	while
break	except	import	or	with
class	false	in	pass	yield
continue	finally	as	raise	
def	for	lambda	return	

A single element that can be named by the symbol is a number. In the following pages, we will see identifiers called certain things like attributes, groups and processes. The following method applies to identifiers.

Rules for Identifiers

At least one character must be used.

An alphabetic letter (upper or lower case) must be the first character, or an underscore, i.e.

- ABCDEFGHIJKLMNOPQRSTUVWXYZabcdefghijklmnopqrstuv wxyz_

The remaining (if any) characters can be alphabetical letters, underscores, or numbers i.e.

- ABCDEFGHIJKLMNOPQRSTUVWXYZabcdefghijklmnopqrstuv wxyz_0123456789

For identifiers, no additional characters (including spaces) are permitted.

Reserved words are disallowed (see Table 2.1).

A few definitions of true and false identifiers are:

Correct identifications are the following terms such that different names may be used: x, x2, sum, port 22, and FLAG.

The first entry (space is not a legal symbol of an identification), four entries (starts with a digit), two (the sign of a pound is not a legal symbol of an identity), and one of the groups (the term is a restricted word).

Zero is a true identification: the subtotal (dash is not the legal sign of an identification).

Python reserves a number of words for special use that could otherwise be used as identifiers. These reserved words or keywords are special and used to define the structure of Python programs and statements. The purposes of many of these reserved words will be revealed throughout this book.

Fortunately, if you accidentally attempt to use one of the reserved words as a variable name within a program, the interpreter will issue an error:

```
>>> class = 23
SyntaxError: invalid syntax
```

Up till now we have avoided keywords completely in our programs. This means there is nothing special about the names print, int, str or type, other than that they happen to be the names of built-in functions. We are free to reassign these names and use them as variables. Consider the following interactive sequence that reassigns the name print to mean something new:

```
>>> print('I am best Robo')
I am best Robo
>>> print
<built-in function print>
>>> type(print)
<class 'builtin_function_or_method'>
>>> print = 100
>>> print
100
>>> print('I am best Robo')
Traceback (most recent call last):
File "<stdin>", line 1, in <module>
TypeError: 'int' object is not callable
type(print)
<class 'int'>
```

Here we used the name print as a variable. In so doing it has lost its original behavior as a function to print the console. While we can reassign the names print, str, type, etc., it generally is not a good idea to do so.

Not only can a function name can be reassigned, but a variable can be assigned to a function:

```
>>> my_print=print
>>> my_print (' hello form my_print!')
hello form my_print!
```

After binding my_print to print we can use my_print in exactly the same way as the built-in print function.

Python is a case-sensitive language. This means that capitalization matters. if is a reserved word, but If, IF or iF are not reserved words. Identifiers also are case sensitive; the variable called Name is different from the variable called name. Note that three of the reserved words (False, None and True) are capitalized.

Variable names should not be distinguished merely by differences in capitalization because this can be confusing to human readers. For the same reason, it is considered poor practice to give a variable the same name as a reserved word with one or more of its letters capitalized.

The most important thing to remember about variables names is that they should be well chosen. A variable's name should reflect the variable's purpose within the program. For example, consider a program controlling a point-of-sale terminal (also known as an electronic cash register). The variable keeping track of the total cost of goods purchased might be named `total` or `total_cost`. Variable names such as `a67_99` and `fred` would be poor choices.

2.4 Various Types of Floating Point Numbers

Many computational tasks require numbers that have fractional parts. For example, to compute the area of a circle given the circle's radius, the value π, or approximately 3.14159, is used. Python supports such noninteger numbers, and they are called floating point numbers. The name implies that during mathematical calculations the decimal point can move or "float" to various positions within the number to maintain the proper number of significant digits. The Python name for the floating-point type is `float`. Consider the following interactive session:

```
>>> a = 45.326
>>> a
45.326
>>> type(a)
<class 'float'>
```

Based on Python's implementation on a particular machine, the set of floating points (positive and negative meaning from small to maximum and minimum) and the specificity of accuracy (the number of available figures). Table 2.2 shows details of the popular 32-bit computer device's floating-point values. All positive and negative numbers may be floating-point numbers that, compared to Python inputs, are arbitrarily large (or inappropriately tiny for negative numbers).

TABLE 2.2

Floating Point Number Attributes for 32-bit Operating Systems

Data Type	Capacity of Storage	Magnitude Lowest	Magnitude Greatest	Minimum precision
float	64 bits	2.22507×10^{-308}	$1.79769 \times 10^{+308}$	15 digits

An approximation of the mathematical value of π is shown in Program 2.5 (pi-print.py).

```
pi = 3.14159;
print ("Pi =", pi)
print ("or", 3.14, "for short")
```

The first line in (pi-print.py) prints the value of the variable `pi`, and the second line prints a literal value. Any literal numeric value with a decimal point in a Python program automatically has the type `float`.

Floating-point numbers are an approximation of mathematical real numbers. The range of floating point numbers is limited, since each value must be stored in a fixed amount of memory. Floating-point numbers differ from integers in another, very important way. Any integer can be represented exactly. This is not true necessarily for a floating-point number. Consider the real number π. π is an irrational number which means it contains an infinite number of digits with no pattern that repeats. Since π contains an infinite number of digits, its value can only be approximated. Because of the limited number of digits available, some numbers with a finite number of digits can be only approximated; for example, the number 45.368954127892 has to be approximated to so many digits of the floating type; Python stores them at 45.368954127892:

```
>>> X = 45.368954127892
>>> X
45.368954127892
```

A case in point is Program 2.6 (valueprint.py) which provides an example of the issues created by the inaccuracy of floating-point numbers.

An example of the problems that can arise due to the inexact nature of floating-point numbers is demonstrated later.

Floating-point numbers can be expressed in scientific notation. Since most programming editors do not provide superscripting and special symbols like ×, the normal scientific notation is altered slightly. The number 6.022×1023 is written as 6.022e23. The number to the left of the e (capital E can be used as well) is the mantissa, and the number to the right of the e is the exponent of 10. As another example, -5.1×10^{-4} is expressed in Python as −5.1e-4. Program 2.7 (scientificnotation.py) prints some scientific constants using scientific notation.

```
avogadros_number = 6.022e23
c = 2.998e8
print ("Avogadro's number =", avogadros_number)
print ("Speed of light =", c)
```

Output

```
Avogadro's number = 6.022e+23
Speed of light = 299800000.0
```

2.5 Control Codes within Strings

The characters that can appear within strings include the letters of the alphabet (A–Z, a–z), digits (0–9), punctuation (. , : etc.), and other printable symbols (#, &, %, etc.). In addition to these "normal" characters, we may embed special characters known as control codes. Control codes control the way text is rendered in a console window or paper printer. The backslash symbol (\) signifies that the character that follows it is a control code, not a literal character. The string '\n' thus contains a single control code. The backslash is known as the escape symbol, and in this case we say the n symbol is escaped. The \n control code represents the new line control code which moves the text cursor down to the next line in the console window. Other control codes include \t for tab, \f for a form feed (or page eject) on a printer, \b for backspace and \a for alert (or bell). The \b and \a do not produce the desired results in the IDLE interactive shell, but they work properly in a command shell. Program 2.8 (specialchars.py) prints some strings containing some of these control codes.

```
print('P\nQ\nR')
print('S\tU\tV')
print('WX\bYZ')
print('11\a 12\a 13\a 14\a 15\a 16')
```

Output

```
P
Q
R
S          U          V
WXYZ
11   12   13   14   15   16
```

On most systems, the computer's speaker beeps five times when printing the last line.

A string with a single quotation mark at the beginning must be terminated with a single quote; similarly, a string with a double quotation mark at the beginning must be terminated with a double quote. A single-quote string may have embedded double quotes, and a double-quote string may have embedded single quotes. If you wish to embed a single quote mark within a single-quote string, you can use the backslash to escape the single quote (\'). An unprotected single quote mark would terminate the string. Similarly, you may protect a double-quote mark in a double-quote string with a backslash (\"). Program 2.9 (escapequotes.py) shows the various ways in which quotation marks may be embedded within string literals.

```
print("Did you know that 'word' is a word?")
print('Did you know that "word" is a word?')
print("Did you know that \'word'\ is a word?")
print("Did you know that \"word\" is a word?")
```

Output

```
Did you know that 'word' is a word?
Did you know that "word" is a word?
Did you know that 'word' is a word?
Did you know that "word" is a word?
```

Since the backslash serves as the escape symbol, in order to embed a literal backslash within a string you must use two backslashes in succession. Program 2.10 (printpath.py) prints a string with embedded backslashes:

```
filename ='C:\\User\\rick'
Print(filename
```

Output

```
C:\Users\rick
```

2.6 User Input

The print function enables a Python program to display textual information to the user. Programs may use the input function to obtain information from the user. The simplest use of the input function assigns a string to a variable:

```
x = input ()
```

The parentheses are empty because the input function does not require any information to do its job. Program 2.11 (usinginput.py) demonstrates that the input function produces a string value:

```
print('Please enter some text:')
x = input()
print('Text entered:', x)
print('Type:', type(x))
```

Output

```
Please enter some text:
12
Text entered: 12
Type: <class 'str'>
```

The second line shown in the output is entered by the user, and the program prints the first, third and fourth lines. After the program prints the message "Please enter some text:", the program's execution stops and waits for the user to type some text using the keyboard. The user can type, backspace to make changes and then type some more. The text the user types is not committed until the user presses the enter (or return) key.

Quite often we want to perform calculations and need to get numbers from the user. The input function produces only strings, but we can use the int function to convert a properly formed string of digits into an integer. Program 2.12 (addintegers.py) shows how to obtain an integer from the user.

```
print('Enter an integer value:')
a = input()
print('Enter another integer value:')
b = input()
val1 = int(a)
val2 = int(b)
print(val1, '+', val2, '=', val1 + val2)
```

Output

```
Enter an integer value:
12
Enter another integer value:
12
12 + 12 = 24
```

Lines two and four represent user input, while the program generates the other lines. The program halts after printing the first line and does not continue until the user provides the input. After the program prints the second message it again pauses to wait for the user's second entry.

Since user input almost always requires a message to the user about the expected input, the input function optionally accepts a string that it prints just before the program stops to wait for the user to respond. The statement

$$x = input('Please enter some text: ')$$

prints the message "Please enter some text:" and then waits to receive the user's input to assign it to x. This can be expressed more compactly using this form of the input function as shown in Program 2.13 (addintegers2.py).

```
x = input ('Please enter an integer value: ')
y = input ('Please enter another integer value: ')
num1 = int (x)
num2 = int (y)
print (num1, '+', num2, '=', num1 + num2)
```

This can be shortened as in Program 2.14 (addnumber3.py). The function's input and int are combined in a single expression.

```
num1 = int (input ('Please enter an integer value: '))
num2 = int (input ('Please enter another integer
value: '))
print (num1, '+', num2, '=', num1 + num2)
```

The expression

```
int (input ('Please enter an integer value: '))
```

uses a practical structure methodology. Rather than utilizing the intermediate variables in Program 2.14 (addintegers2.py), the output of the input function is transferred directly to the int method. The functional design is also used to improve our software code.

2.7 Evaluation (eval( )) Function

The `input` function produces a string from the user's keyboard input. If we wish to treat that input as a number, we can use the `int` or `float` function to make the necessary conversion:

```
x = float(input('Please enter a number'))
```

Here, whether the user enters 2 or 2.0, x will be a variable with the type floating point. What if we wish x to be of type integer if the user enters 2 and x to be a floating point if the user enters 2.0? Python provides the `eval` function that attempts to evaluate a string in the same way that the interactive shell would evaluate it. Program 2.15 (evalfunc.py) illustrates the use of `eval`.

```
x1 = eval(input('Entry x1? '))
print('x1 =', x1, ' type:', type(x1))

x2 = eval(input('Entry x2? '))
print('x2 =', x2, ' type:', type(x2))

x3 = eval(input('Entry x3? '))
print('x3 =', x3, ' type:', type(x3))

x4 = eval(input('Entry x4? '))
print('x4 =', x4, ' type:', type(x4))

x5 = eval(input('Entry x5? '))
print('x5 =', x5, ' type:', type(x5))
```

Output

```
Entry x1? 4
x1 = 4 type: <class 'int'>
Entry x2? 4.0
x2 = 4.0 type: <class 'float'>
Entry x3? 'x1'
x3 = 4 type: <class 'int'>
Entry x4? 4
x4 = 4 type: <class 'int'>
Entry x5? x6?x6
```

Notice that when the user enters 4, the variable's type is integer. When the user enters 4.0, the variable is a floating-point variable. For x3, the user supplies the string `'x3 '` (note the quotes), and the variable's type is `string`. The more interesting situation is x4. The user enters x1 (no quotes). The eval function evaluates the non-quoted text as a reference to the name x1. The program binds the name x1 to the value 4 when executing the first line of the program. Finally, the user enters x6 (no quotes). Since the quotes are missing, the eval function does not interpret x6 as a literal string; instead it treats x6 as a name an attempts to evaluate it. Since no variable named x6 exists, the eval function prints an error message.

The eval function dynamically translates the text provided by the user into an executable form that the program can process. This allows users to provide input in a variety of flexible ways; for example, users can enter multiple entries separated by commas, and the eval function evaluates it as a Python tuple. This allows for tuple assignment, as seen in Program 2.16 (addnumber 3.py).

```
num1 = eval(input('Please enter number : '))
num2 = eval(input('Please enter number 2: '))
print(num1, '+', num2, '=', num1 + num2)
```

The illustration below shows how the user now must reach the two numbers separated by a comma at the same time:

```
Please enter number 1, number 2: 23, 10
23+10 = 33
```

As the above is like an empty dynamic container, of one line of basic Python script, only one phrase is allowed by the user. It is not a single word.

```
print(eval(input()))
```

Program 2.17 (entervalue.py) indicates that users can input an arithmetic expression and handle it correctly. The output is:

```
5 + 20
25
```

The user inserts the text 5 + 20 and prints 25 on the application. Notice that the program (entervalue.py) is not foreseen in the introduction; the system performs evaluable functions, then compiles and executes the provided text to build 25.

2.8 Controlling (print( )) Function

In Program 2.12 (addintegers.py) we prefer that the cursor remains at the end of the printed line so when the user types a value it also appears on the same line as the message prompting for the values. When the user presses the enter key to complete the input, the cursor automatically moves down to the next line. The print function as we have seen so far always prints a line of text, and then the cursor moves down to the next line so any future printing appears on the next line. The print statement can accept an additional argument that allows the cursor to remain on the same line as the printed text:

```
print('Please enter an integer value:', end=")
```

The expression end=" is known as a keyword argument. The term keyword here means something different from that used to mean a reserved word. We defer a complete explanation of keyword arguments until we have explored more of the Python language. For now it is sufficient to know that a print function call of this form will cause the cursor to remain on the same line as the printed text. Without this keyword argument, the cursor moves down to the next line after printing the text.

The print statement

```
print('Please enter an integer value: ', end=")
```

means: Print the message "Please enter an integer value:", and then terminate the line with nothing, rather than the normal \n newline code. Another way to achieve the same result is:

```
print(end='Please enter an integer value: ')
```

This statement means: Print nothing, and then terminate the line with the string "Please enter an integer value:" rather than the normal \n newline code. The behavior of the two statements is indistinguishable.

The statement

```
print('Please enter an integer value:')
```

is an abbreviated form of the statement

```
print('Please enter an integer value:', end='\n')
```

that is, the default ending for a line of printed text is the string '\n', the newline control code. Similarly, the statement

```
print()
```

is a shorter way to express

```
print(end='\n')
```

Observe closely the output of Program 2.18 (printingexample.py).

```
print('A', end='')
print('B', end='')
print('C', end='')
print()
print('X')
print('Y')
print('Z')
```

Output

```
ABC
X
Y
Z
```

The statement

```
print()
```

essentially moves the cursor down to next line.

Sometimes it is convenient to divide the output of a single line of printed text over several Python statements. As an example, we may want to compute part of a complicated calculation, print an intermediate result, finish the calculation, and print the final answer with the output all appearing on one line of text. The end keyword argument allows us to do so.

Another keyword argument allows us to control how the print function visually separates the arguments it displays. By default, the print function places a single space in between the items it prints. print uses a keyword argument named sep to specify the string to be inserted between items. The name sep stands for separator. The default value of sep is the string ' ', a string containing a single space. Program 2.19 (countong.py) shows the sep keyword customizes print's behavior.

```
w, x, y, z = 10, 15, 20, 25
print(w, x, y, z)
print(w, x, y, z, sep=',')
print(w, x, y, z, sep='')
print(w, x, y, z, sep=':')
print(w, x, y, z, sep='-----')
```

Output

```
10  15  20  25
10,  15,  20,  25
10152025
10:15:20:25
10-----15-----20-----25
```

The first part of the output shows `print`'s default method of using a single space between printed items. The second output line uses commas as separators. The third line runs the items together with an empty string separator. The fifth line shows that the separating string may consist of multiple characters.

2.9 Conclusion

Data type in Python is a crucial term in programming. Variables can store a variety of data, and different data types can perform different tasks. A data type, also referred to as `type`, is a property of data that informs the interpreter or compiler about how the programmer intends to use the available information. The categorization or grouping of data objects is known as data forms. This denotes the type of value that determines which operations can be performed on a given collection of data. Data types are simply classes, and variables are instances (objects) of these classes, since everything in Python programming is an entity.

3

Operators

3.1 Python Expressions and Operators

A literal value like 34 and a variable like x are examples of simple expressions. Values and variables can be combined with operators to form more complex expressions. In Section 2.1 we saw how we can use the + operator to add integers and concatenate strings. Program 3.1 (addition.py) shows how the addition operator (+) can used to add two integers provided by the user [11].

```
num1 = eval(input('Enter the First number: '))
num2 = eval(input('Enter the Second number: '))
sum = num1 + num2
print(num1, '+', num2, '=', sum)
```

Output

```
Enter First number: 12
Enter Second number: 1
  12 + 1 = 13
```

```
        num1 = eval(input('Enter First number: '))
```

This statement prompts the user to enter some information. After displaying the prompt string "Please enter an integer value:", this statement causes the program's execution to stop and wait for the user to type in some text and then press the enter key. The string produced by the input function is passed to the eval function which produces a value to assign to the variable value1. If the user types the sequence 431 and then presses the enter key, value1 is assigned the integer 431. If instead the user enters 23 + 23, the variable is assigned the value 46

```
        num2 = eval(input('Enter the Second number: '))
```

DOI: 10.1201/9781003202035-3

This statement is identical to the first.

```
sum = num1 + num2
```

This is an assignment statement because it contains the assignment operator (=). The variable `sum` appears to the left of the assignment operator, so `sum` will receive a value when this statement executes. To the right of the assignment operator is an arithmetic expression involving two variables and the addition operator [12]. The expression is evaluated by adding together the values bound to the two variables. Once the addition expression's value has been determined, that value can be assigned to the `sum` variable.

```
print(num1, '+', num2, '=', sum)
```

All expressions have a value. The process of determining the expression's value is called evaluation. Evaluating simple expressions is easy. The literal value 54 evaluates to 54. The value of a variable named x is the value stored in the memory location bound to x. The value of a more complex expression is found by evaluating the smaller expressions that make it up and combining them with operators to form potentially new values.

The commonly used Python arithmetic operators are found in Table 3.1. These operations – addition, subtraction, multiplication, division and power – behave in the expected way. The `//` and `%` operators are not common arithmetic operators in everyday practice, but they are very useful in programming. The `//` operator is called integer division, and the `%` operator is the modulus or remainder operator. 25/3 is 8.3333. Three does not divide into 25 evenly. In fact, three goes into 25 eight times with a remainder of one. Here, eight is the quotient, and one is the remainder. 25//3 is 8 (the quotient), and 25%3 is 1 (the remainder).

TABLE 3.1

Simple Generator Python Arithmetic

Python Expression	Description
$x+y$	x started adding to y, if y is x and y x if x and y are strings x and y
$x-y$	x take y if there are numbers x and y
$x*y$	x times y when x and y are numbers x are y times, when x is a string and y is a y integer, when y is a string and x is an integer, x times
x/y	if x and y are the numbers, x is divided by y
$x//y$	floor of x/y, if x and y are numbers
$x\%y$	Otherwise, if x and y are numbers, x is divided by y
$x**y$	if x and y are numbers, x raised to y power

All these operators are classified as binary operators because they operate on two operands. In the statement

```
x = y + z;
```

on the right side of the assignment operator is an addition expression y + z. The two operands of the + operator are y and z.

Two operators, + and -, can be used as unary operators. A unary operator has only one operand. The unary operator expects a single numeric expression (literal number, variable or more complicated numeric expression within parentheses) immediately to its right; it computes the additive inverse of its operand. If the operand is positive (greater than zero), the result is a negative value of the same magnitude; if the operand is negative (less than zero), the result is a positive value of the same magnitude. Zero is unaffected [13].

Write code to print the program given below.

```
x, y, z = 3, -4, 0
x = -x
y = -y
z = -z
print(x, y, z)
```

Output

-7 9 0

The statement is:

```
>>>print(-(4 - 5))
```

Output

1

The unary + operator is present only for completeness; when applied to a numeric value, variable or expression, the resulting value is no different from the original value of its operand. Omitting the unary + operator from the following statement

```
x = +y
```

does not change its behavior.

All the arithmetic operators are subject to the limitations of the data types on which they operate; for example, consider the following interaction sequence:

```
>>> 3.0**10
59049.0
>>> 3.0**100
5.153775207320113e+47
>>> 2.0**1000
1.0715086071862673e+301
>>> 3.0**10000
Traceback (most recent call last):
File "<stdin>", line 1, in <module>
OverflowError: (34, 'Result too large')
```

The 2.0**1000 definition does not determine the correct response since the correct answer comes beyond the floating-point range.

The product is an integer for the +, -, *, /, %, or ** operators. The operator added to two integers produces the product of a floating point, as in the statement:

```
print(10/3, 3/10, 10//3, 3//10)
3.3333333333333335 0.3 3 0
```

The first two results are the same as a hand-held calculator would produce. The second two results use integer division, and 3 goes into 10 three times, while 10 goes into 3 zero times. The // operator produces an integer result when used with integers because in the first case 10 divided by 3 is 3 with a remainder of 1, and in the second case 3 divided by 10 is 0 with a remainder of 3. Since integers are whole numbers, any fractional part of the answer must be discarded. The process of discarding the fractional part leaving only the whole number part is called truncation. Truncation is not rounding; for example, 11/3 is 3.6666..., but 11//3 truncates to 3. Be warned: truncation simply removes any fractional part of the value.

It does not round. Both 10.01 and 10.999 truncate to 10.

The modulus operator (%) computes the remainder of integer division; thus in

```
print(20%3, 3%20)
```

the output is

```
2 3
```

since 20 divided by 3 is 3 with a remainder of 2, and 3 divided by 10 is 0 with a remainder of 3.

The modulus operator is more useful than it may first appear. Program 3.2 (timeconv.py) shows how it can be used to convert a given number of seconds to hours, minutes and seconds.

Floating-point arithmetic always produces a floating-point result.

```
print(20.0/3.0, 3.0/20.0, 20.0//3.0, 3//20.0)
```

Output

```
6.666666666666667 0.15 6.0 0.0
```

Recall from Section 2.4 that integers can be represented exactly, but floating-point numbers are imprecise approximations of real numbers. Program 3.2 (cal.py) clearly demonstrates the weakness of floating point numbers.

```
one = 1.0
one_third = 1.0/3.0
zero = one - one_third - one_third - one_third
print('one =', one, ' one_third =', one_third, ' zero =',
zero)
```

Output

```
one = 1.0 one_third = 0.3333333333333333 zero =
1.1102230246251565e-16
```

The reported result is $1.1102230246251565 \times 10^{-16}$, or 0.0000000000000001110 2230246251565. While this number is very small, with real numbers we get

$$1 - \frac{1}{3} - \frac{1}{3} - \frac{1}{3} = 0$$

Floating-point numbers are not real numbers, so the result of 1.0/3.0 cannot be represented exactly without infinite precision. In the decimal (base 10) number system, one-third is a repeating fraction, so it has an infinite number of digits. Even simple non-repeating decimal numbers can be a problem. One-tenth (0.1) is obviously non-repeating, so it can be expressed exactly with a finite number of digits. As it turns out, since numbers within computers are stored in binary (base 2) form, even one-tenth cannot be represented exactly with floating-point numbers, as Program 3.3 (Call.py) shows.

```
one = 1.0
one_tenth = 1.0/10.0
zero = one - one_tenth - one_tenth - one_tenth \
          - one_tenth - one_tenth - one_tenth \
          - one_tenth - one_tenth - one_tenth \
          - one_tenth

print('one =', one, ' one_tenth =', one_tenth, ' zero =',
zero)
```

Output

```
one = 1.0  one_tenth = 0.1  zero = 1.3877787807814457e-16
```

Probably, the response that has been stated is similar to the right answer (1.38777877807814457 number 10^{-15}). It is correct at the one hundred trillionth position (15 positions behind the decimal point).

When this subtraction of a single statement had been written on one side, it will float across or off the edit pane. Typically, at the end of the source code section, a Python declaration stops. At the end of an unfinished line, a programmer may split it into two or more lines with the symbol backslash (\). The translator then enters the next line as the line finishes with \. And a very long but full Python declaration is used by the interpreter.

Despite their inaccuracy, floating points are used to address complex research and technological challenges every day across the globe. The drawbacks of floating-point numbers are inherent since the properties of infinite features cannot be interpreted finitely. However, floating-point numbers do give a reasonable degree of practical accuracy.

Mixed elements may include expressions; for instance, in the following program fragment

```
x = 4
y = 10.2
z sum = x + y
```

x is a whole and y is a floating-point number. What is the shape of the term x + y? Arithmetical expressions are created and saved for the operator only with integer values. Both arithmetical operators perform floating-point checks. If one operator uses mixed operators – one runs an integer and the other a floating-point number – then there is a floating-spot value for an interpreter to represent the integer function. This assumes x + y is a fixed point function and a floating-point attribute is allocated as the number.

3.1.1 Comparison Operators

TABLE 3.2

Comparison Operators

Symbol of Operator	Combiniation of Operator and Operand	Description	Output
==	a==b	Equal to	True if the value of a is equal to the value of b False otherwise
!=	a!=b	Not equal to	True if a is not equal to b False otherwise
<	a<b	Less than	True if a is less than b False otherwise
<=	a<=b	Less than or equal to	True if a is less than or equal to b False otherwise
>	a>b	Greater than	True if a is greater than b False otherwise
>=	a>=b	Greater than or equal to	True if a is greater than or equal to b False otherwise

Examples of the operators in use are:

```
>>> a = 40
>>> b = 30
>>> a == b
False
>>> a != b
True
>>> a <= b
True
>>> a >= b
False

>>> a = 50
>>> b = 50
>>> a == b
True
```

```
>>> a <= b
True
>>> a >= b
True
```

Boolean cases, such as conditional and loop statements for program flow, as we will see later, are usually used by comparison operators.

3.1.1.1 Floating-Point Value Equality Comparison

Remember from the previous conversation on the number of floating points that it may not be exactly that value which was stored internally for the entity of a floating point number. Therefore, the comparison of floating points with exact equality is a bad technique [14]. Take the example:

```
>>> x = 4.4 + 5.5
>>> x == 9.9
False
```

The internal representations of the supplementary operand are not exactly 4.4 and 5.5, so you cannot depend on x to equate precisely to 9.9.

The preferred approach to decide whether two floating points are "similar" is to calculate if the tolerances are identical to each other:

```
>>> tolerance = 0.00001
>>> x = 1.1 + 2.2
>>> abs(x - 3.3) < tolerance
True
```

abs () returns the absolute value. If the relative value of the variance is lower than the given number, they are sufficiently similar to each other.

3.1.2 Logical Operators

Logical operators do not use expressions evaluated in Boolean expressions to construct dynamic conditions, or to alter and combine them.

3.1.2.1 *Logical Expressions Using Boolean Operands*

As you can see, some of the objects and Python expressions are Boolean. In other words, one of the true or false objects in Python is identical. Take the following examples:

```
>>> x = 15
>>> x < 30
True
>>> type(x <30)
<class 'bool'>

>>> r = x > 30
>>> r
False
>>> type(r)
<class 'bool'>

>>> callable(y)
False
>>> type(callable(y))
<class 'bool'>

>>> r = callable(len)
>>> r
True
>>> type(r)
<class 'bool'>
```

In these examples, all the Boolean objects or expressions are x < 30 and callable(y). The interpretation of logical statements which do not include or use simple Boolean expressions:

TABLE 3.3

Boolean Values

Boolean Operator	Example	Description
not	not x	If x is false, it is true
		If x is true, it is false
		(Returns the context of x logically)
or	x or y	It's true if x or y are true
		false
and	x and y	True if you are both x and y
		otherwise false

Take a look at how they work in practice below.
"not" and Boolean Operands

```
>>>x = 5
>>>not x < 10
False
>>>not callable(x)
True
```

"or" and Boolean Operands

```
>>>x = 5
>>>x < 10 or callable(x)
True
>>>x < 0 or callable(x)
False
```

"and" and Boolean Operands

```
>>>x = 5
>>>x < 10 and callable(x)
False
>>>x < 10 and callable(len)
True
```

3.1.2.2 Evaluation of Boolean and Non-Boolean Expressions

Many objects and statements are not true or false. They can also be assessed and calculated to be "truthful" or "falsified" in a Boolean expression, though.

This is well described in Python. In Boolean terms, all of the following are considered false:

- A False Boolean value;
- Every numerically null value (0, 0.0, 0.0+0.0j);
- A null string;
- An object with an empty composite data form;
- A special Python keyword meaning "none."

Virtually any other Python-built object is considered valid.

With the integrated bool() function, you can define the "truthfulness" of an object or expression. A bool() return statement is true or false [15].

- **Python Numeric Value**

  A value of zero is false.

  A value that is non-zero is true.

```
>>> print(bool(0), bool(0.0), bool(0.0+0j)
False False False

>>> print(bool(-3), bool(3.14159), bool(1.0+1j)
True True True
```

- **Python String**

  A string that is empty is false.

  A string that is not empty is true.

```
>>> print(bool("), bool(""), bool(""""""")
False False False

>>> print(bool('foo'), bool(" "), bool("' "')
True True True
```

- **Composite Data Object Built-In Function**

  Python offers built-in, array, tuple and dict composite data types. There are types of "containers" for other items. An object of one such kind is called an attribute if it is vacant and right if it is not vacant.

  The following explanations show the form of collection. (Lists with square brackets are described in Python.)

  For example in the list, tuple, dict and set are the types.

```
>>> type ([])
<class 'list'>
>>> bool ([])
False
>>> type ([1,2,3])
<class 'list'>
>>> bool ([1,2,3])
True
```

- **The "None" Keyword**

  None is always false:

  ```
  >>> bool(None)
  False
  ```

- **Non-Boolean Operand Expressions**

  Non-Boolean variables can be configured and added with either not or and. The outcome is dependent on the operand's "truthfulness."

  "not" and Non-Boolean Operands

  For a non-Boolean value, x, this is what happens:

  ```
  >>> x = 4
  >>> bool(x)
  True
  >>> not x
  False
  >>> x = 0.0
  >>> bool(x)
  False
  >>> not x
  True
  ```

  "or" and Non-Boolean Operands

  For two non-Boolean values x and y the following happens.

  In this case, the expression x or y does not know either truth or false, but only one expression x or y:

  ```
  >>> x = 3
  >>> y = 4
  >>> x or y
  3

  >>> x = 0.0
  >>> y = 4.4
  >>> x or y
  4.4
  ```

This is always the case: if x or y are true and if both x and y are false, the expression x or y is true.

"and" and Non-Boolean Operands

For the two non-Boolean values x and y, here is what you will receive:

```
>>> x = 3
>>> y = 4
>>> x and y
4
>>> x = 0.0
>>> y = 4.4
>>> x and y
0.0
```

Check the x and y values.

- **Logical Expressions of Compounds and Evaluation of Short Circuits**

So far only a single operator is used between two operands:

```
x or y
x and y
```

The combined compound logical expressions of multiple logical operators and operators are created.

Compound "or" expressions

Take the following statement:

```
x₁ or x₂ or x₃ or ... xₙ
```

If all of the xi are valid, this expression is true.

- o In a term such as this, Python uses the technique known as short-course assessment in honor of computer scientist John McCarthy, also known as McCarthy assessment. From left to right the xi operands are assessed. The whole phrase is understood to be true until one is proved to be true. Python fails at that stage and no longer evaluates conditions. The value of the whole word is the value of the xi which ends the evaluation.

o To assist in the evaluation of the shortcut, suppose that the following is a simple 'identity' function: f():

o f() takes one point.

o The argument is returned to the console.

o The reason forwarded to the console is returned when the argument returns.

(In the upcoming functions guide, you can see how to describe this function.)

Below are some instances of calls to f():

```
>>> f(0)
-> f(0) = 0
0

>>> f(False)
-> f(False) = False
False

>>> f(1.5)
-> f(1.5) = 1.5
1.5
```

Because f() only returns the passed statement, the expression is only true or false. Furthermore, f() shows the statement on the console and checks whether it was named or not.

Now take the following logical expression into account:

```
>>> f(0) or f(False) or f(1) or f(2) or f(3)
-> f(0) = 0
-> f(False) = False
-> f(1) = 1
1
```

The first interpreter assesses f(0), which is 0. A value of 0 is incorrect. The phrase is not yet accurate, because the assessment needs to be correct. False returns the next operand, f(False). This is indeed wrong, but the assessment begins.

Then it's f(1). This is assessed at 1, which is valid. The interpreter stops at that moment, for now the whole phrase is known to be valid. 1 is returned as the expression value and neither f(2) nor f(3) are evaluated for the remainder of the operands. The f(2) and f(3) calls are not seen in the window.

- **Preventing an Exception**

  If two variables a and b are described, you would like to know if (b/a) > 0:

```
>>> a = 4
>>> b = 1
>>> (b/a) > 0
True
```

There is the chance of a 0, in which case an exception is raised by the interpreter:

```
>>> a = 0
>>> b = 1
>>> (b/a) > 0
Traceback (most recent call last):
    File "<pyshell#2>", line 1, in <module>
        (b/a) > 0
ZeroDivisionError: division by zero
```

An error in such an expression should be avoided:

```
>>> a = 0
>>> b = 1
>>> a != 0 and (b/a) > 0
False
```

If a is 0 = 0 it is wrong. The same assessment ends.

(b/a) is not executed and there is no error in this expression. If the number is real, it is very precise. When a is 0, the expression is false. An explicit analogy is not necessary a!=0:

```
>>> a = 0
>>> b = 1
>>> a and (b/a) > 0
0
```

- **Default Value Selection**

    Another value selection is that when a set value is 0, or null, it is a default value. For instance, suppose you want to add a variable s to the value of a string. But you want to have a default value when the string is zero.

    A short-circuit assessment is a succinct way for this to be expressed:

$$s = \text{"<default value>"} \quad or \quad string$$

If string is non-empty, it's valid, and it's true at this stage of the expression string, or is '<default worth>.' Assessment ends and the string value is returned and allocated to s:

```
>>> string = 'foo bar'
>>> s = string or '<default_value>'
>>> s
'foo bar'
```

On the other hand, if a string is an empty string, it is false. Evaluation of the string or '<default_value>' continues to the next operand, '<default_value>', which is returned and assigned to s:

```
>>> string = ''
>>> s = string or '<default_value>'
>>> s
'<default_value>'
```

3.1.3 Chained Comparisons

Operators of comparison may be chained to arbitrary lengths. The following terms, for example, are almost equal:

```
x < y <= z
x < y and y <= z
```

Both are assessed at the same Boolean value. The delicate distinction between the two is that y is assessed only once in the chained relation x < y < = z. Y would be tested twice for the longer expression x < y and y < = z.

This would not make an important distinction in situations where Y is a static attribute. But take into consideration these terms:

```
x < f() <= z
x < f() and f() <= z
```

If f() is a function which modifies program data, it is important to distinguish between it, once in the first case and twice in the second case.

If op1, op2,... opn are comparative operators in general, then the following are Boolean with the same meaning:

$$x_1 \; op_1 \; x_2 \; op_2 \; x_3 \; ... \; xn_{-1} \; op_n \; x_n$$
$$x_1 \; op_1 \; x_2 \; and \; x_2 \; op_2 \; x_3 \; and \; ... \; xn_{-1} \; op_n \; x_n$$

Every xi is evaluated only once in the former case. In the above scenario, all are assessed twice, excluding the first and last one, unless a short-circuit assessment triggers premature completion.

3.1.4 Bitwise Operators

Bit by bit means bitwise; operators treat operands as binary digit sequences and work bit by bit on them.

TABLE 3.4

Bitwise Operators

Symbol of Operator	Expression	Description	Output
&	a&b	bitwise AND	The logical AND of the bits in the corresponding position of the operands in any bit position in the result. (1 if they are both 1, otherwise they are 0.)
|	a|b	bitwise OR	The logical OR of the bits in the corresponding position of the operands are each bit position in the consequence. (1 if one is one, another 0.)
~	~a	bitwise negation	The logical negation of the bit in the corresponding operand position is the position of each bit in the outcome. (1 in case 0, 0 in case 1.)

(Continued)

TABLE 3.4 (CONTINUED)

Bitwise Operators

Symbol of Operator	Expression	Description	Output
^	a^b	bitwise XOR (exclusive OR)	The logical XOR for the bits in the corresponding position of the operands is each bit position inside the result. (1 if there are differences in the operands, 0 if they are identical).
>>	a>>n	shift right n places	Each bit is shifted right n places
<<	a<<n	shift left n places	Each bit is shifted left n places

Here are some examples:

```
Python 3.9.6 (tags/v3.9.6:db3ff76, Jun 28 2021, 15:26:21)
[MSC v.1929 64 bit (AM D64)] on Win32
Type "help", "copyright", "credits" or "license()" for
more information.
>>> a=60
>>> b=13
>>> c=a&b
>>> c
12
>>> a | b
61
>>> ~a
-61
>>> a^b
49
>>> a>>n
Traceback (most recent call last):
    File "<pyshell#7>", line 1, in <module>
        a>>n
NameError: name 'n' is not defined
>>> a>>b
0
>>> a<<b
491520
>>>
```

The intention of the.format() '0b{:04b}' is to lay out the numerical output to facilitate the reading of the bit operations. Much more information will be provided later about the format() process. Pay heed to the operators and the performance of bit-by- bit operations.

3.1.5 Identity Operators

Python offers two operators that decide and do not decide if the operand is the same – that is, refers to the same thing [16]. The two operands apply to objects which contain the same data but are not exactly the same entity.

```
>>> x = 1001
>>> y = 1000 + 1
>>> print(x,y)
1001 1001

>>> x == y
True
>>> x is y
False
```

In this case, both x and y refer to items of 1001 weight. They are the same. But you should check the value does not belong to the same objects:

```
>>> id(x)
60307920
>>> id(y)
60307936
```

x and y are not identical, and comparing x and y variables returns a null value.

We've seen before that Python produces only a second reference to the same object while doing a task like x = y and that with the id() function you can prove that. It can also be verified by the operator:

```
>>> a = 'I am a string'
>>> b = a
>>> id(a)
55993992
>>> id(b)
55993992
>>> a is b
True
>>> a == b
True
```

Because the same entity is referenced by a and b, it is reasonable for a and b to be the same.

It is not surprising that the reverse is not the case:

```
>>> x = 10
>>> y = 20
>>> x is not y
True
```

3.1.6 Augmented Assignment Operators

We have also seen that the vector is assigned with a single equal sign (=). The correct meaning of the assignment is of course perfectly viable as an expression that contains other variables:

```
>>> a = 10
>>> b = 20
>>> c = a * 5 + b
>>> c
70
```

Indeed, references to the vector delegated can be found in the phrase to the right of the assignment:

```
>>> a = 10
>>> a = a + 5
>>> a
15

>>> b = 20
>>> b = b * 3
>>> b
60
```

In the above example in the first case 5 is added then the actual value of a is 15. In the second case the actual value of b is 20 and multiplied by 3 it increases three times.

Naturally, this kind of task only makes perfect sense if a value was previously allocated to the attribute concerned:

```
>>> z = z/12
Traceback (most recent call last):
    File "<pyshell#11>", line 1, in <module>
        z = z/12 NameError: name 'z' is not defined
```

3.1.7 Comparison of Arithmetic and Bitwise Operators

For these arithmetic and bitwise operators, Python supports a symbol notation:

TABLE 3.5

Augmented Assignment Notation

Arithmetic Operator	Bitwise Operator
+	
−	&
*	|
/	^
%	>>
//	<<
**	

As per the above table, the operator syntax is used as follows:

```
x <op> = y
x = x <op> y
```

3.2 Operator Associativity and Precedence

When different operators appear in the same expression, the normal rules of arithmetic apply. All Python operators have a precedence and associativity:

- Precedence: when an expression contains two different kinds of operators, which should be applied first?

- Associativity: when an expression contains two operators with the same precedence, which should be applied first?

To see how precedence works, consider the expression

```
>>> 45 + 5 * 20
145
```

Ambiguity exists here. Can Python first add 20 + 4 and then increase the number to 10? Or should 4 * 10 be multiplied in the first place, and 20 then added? Of course, Python has decided on the latter for 60; if the former had been selected, the outcome would be 240. This is a regular algebraic operation, which is used in almost all languages.

Both operators supported by the language are given priority. The operators with the highest precedence are exercised first in one term. Upon obtaining these results, the next highest precedent is processed. This continues until the whole expression is evaluated. When operators are of equal priority they are executed in in left-to-right order.

The Python operators we have seen thus far are from the lowest to the highest in order of precedence.

TABLE 3.6

Operator Associativity and Precedence

	Symbol of Operator	**Description**
Lowest precedence	or	Boolean OR
	and	Boolean AND
	not	Boolean NOT
	==, !=, <, <=, >, >=, is, is not	comparison, identity
	|	bitwise OR
	^	bitwise XOR
	&	bitwise AND
	<<, >>	bit shifts
	+, −	addition, subtraction
	*, /, //, %	mutliplication, division, floor division, modulo
	+x, −x, ~x	unary positive, unary negation, bitwise negation
Highest precedence	**	exponentiation

Every operator in the same table row has the same precedent.

Similarly, in the following case, 3 is lifted to a power level of 4 (81) and then multiplied from left to right (2* 81 * 5 = 810):

```
>>> 2 * 3 ** 4 * 5
810
```

Precedence of the operator can be overridden using parentheses. Parenthesis sentences are always executed first. This happens as follows:

```
>>> 20 + 4 * 10
60
>>> (20 + 4) * 10
240
>>> 2 * 3 ** 4 * 5
810
>>> 2 * 3 ** (4 * 5)
6973568802
```

In the first instance, 20 + 4 is calculated, then 10. In the second instance, 4 * 5, first, then 3;, the result is multiplied by 2, then 3 is calculated.

Nothing is incorrect, even though they are not required to alter the assessment order, with generous use of parentheses. In reality, this is considered good practice because it makes code easier to read and makes clear the order of precedence. Consider the following:

```
(a < 10) and (b > 30)
```

The brackets here are completely redundant because the operators' preferences dominate and so are executed first. However, some may think that the use of the parenthesized version makes things clearer:

```
a < 10 and b > 30
```

If you think that it makes the code more readable, you can still use brackets, even if they are not mandatory for evaluation order.

The expression below shows operator precedence functionality

```
2 + 3 * 4
```

Should it be interpreted as

```
(2 + 3) * 4
```

(that is, 20), or rather as

```
2 + (3 * 4)
```

(that is, 14)? As in normal arithmetic, multiplication and division in Python have equal importance and are performed before addition and subtraction. In the expression

```
2 + 3 * 4
```

the multiplication is performed before addition. The result is 14. The multiplicative operators (*, /, // and %) have equal precedence with each other, and the additive operators (binary + and -) have equal precedence with each other. The multiplicative operators have precedence over the additive operators.

As in standard arithmetic, in Python if the addition is to be performed first, parentheses can override the precedence rules. The expression

```
(2 + 3) * 4
```

evaluates to 20. Multiple sets of parentheses can be arranged and nested in anyway that is acceptable in standard arithmetic.

To see how associativity works, consider the expression

```
2 - 3 - 4
```

The operators in each row have a higher precedence than the operators below it. Operators within a row have the same precedence.

The two operators are the same, so they have equal precedence. Should the first subtraction operator be applied before the second, as in

```
(2 - 3) - 4
```

(that is, −5), or rather is

```
2 - (3 - 4)
```

(that is, 3) the correct interpretation? The former (−5) is the correct interpretation. We say that the subtraction operator is left associative, and the evaluation is left to right. This interpretation agrees with standard arithmetic rules. All binary operators except assignment are left associative.

The assignment operator supports a technique known as chained assignment. The code

```
w = x = y = z
```

should be read right to left. First y gets the value of z, then y gets z's value as well. Both w and x get z's value also.

As in the case of precedence, parentheses can be used to override the natural associativity within an expression.

The unary operators have a higher precedence than the binary operators, and are right associative. This means the statements

```
print (-5 + 4) print (- (5 + 4))
```

have the output

```
-1
-5
```

3.3 Comments in Python Programming

Good programmers annotate their code by inserting remarks that explain the purpose of a section of code or why they chose to write it the way they did. These notes are meant for human readers, not the interpreter. It is common in industry for programs to be reviewed for correctness by other programmers or technical managers. Well-chosen identifiers (see Section 2.3) and comments can aid this assessment process. Also, in practice, teams of programmers develop software. A different programmer may be required to finish or fix a part of the program written by someone else. Well-written comments can help others understand new code more quickly and increase their productivity when modifying old or unfinished code. While it may seem difficult to believe, even the same programmer working on his or her own code months later can have a difficult time remembering what various parts do. Comments can help greatly.

Any text contained within comments is ignored by the Python interpreter. The # symbol begins a comment in the source code. The comment continues in effect until the end of the line:

```
# Compute the average of the values avg = sum / number
```

The first line here is a comment that explains what the statement that follows it is supposed to do. The comment begins with the # symbol and continues until the end of that line. The interpreter will ignore the # symbol and the contents of the rest of the line. You also may append a short comment to the end of a statement:

```
avg = sum / number # Compute the average of the values
```

Here, an executable statement and the comment appear on the same line. The interpreter will read the assignment statement, but it will ignore the comment.

How are comments best used? Avoid making a remark about obvious things; for example:

```
result = 0 # Assign the value zero to the variable named result
```

The effect of this statement is clear to anyone with even minimal Python programming experience. Thus, the audience of the comments should be taken into account; generally, "routine" activities require no remarks. Even though the effect of the above statement is clear, its purpose may need a comment. For example:

```
result = 0 # Ensures 'result' has a well-defined minimum value
```

This remark may be crucial for readers to completely understand how a particular part of a program works. In general, programmers are not prone to providing too many comments. When in doubt, add a remark. The extra time it takes to write good comments is well worth the effort.

3.4 Bugs in Programs

Beginner programmers make mistakes writing programs because of inexperience in programming in general or because of unfamiliarity with a programming language. Seasoned programmers make mistakes due to carelessness or because the proposed solution to a problem is faulty and the correct implementation of an incorrect solution will not produce a correct program [17].

In Python, there are three general kinds of errors: syntax errors, run-time errors and logic errors.

3.4.1 Syntax Errors

The interpreter is designed to execute all valid Python code. The interpreter reads the Python source code and translates it into executable machine code.

This is the translation phase. If the interpreter detects an invalid program during the translation phase, it will terminate the program's execution and report an error. Such errors result from the programmer's misuse of the language. A syntax error is a common error that the interpreter can detect when attempting to translate a Python statement into machine language. For example, in English one can say

> The boy goes quickly.

The correct syntax is used in this paragraph. However, the paragraph

> The kid goes hard.

is not correct syntactically: the number of the subject (singular form) disagrees with the number of the verb (plural form). It contains a syntax error. It violates a grammatical rule of the English language. Similarly, the Python statement

```
x = y + 2
```

is syntactically correct because it obeys the rules for the structure of an assignment statement described in Section 2.2. However, consider replacing this assignment statement with a slightly modified version:

```
y + 2 = x
```

If a statement like this one appears in a program, the interpreter will issue an error message; for example, if the statement appears on line 12 of an otherwise correct Python program described in a file named error.py, the interpreter reports:

```
>>> x + 2 = y
File "error.py", line 12
SyntaxError: can't assign to operator
```

The syntax of Python does not allow an expression like y + 2 to appear on the left side of the assignment operator.

Other common syntax errors arise from simple typographical errors like mismatched parentheses or string quotes or faulty indentation.

3.4.2 Run-time Errors

A syntactically correct Python program can still have problems. Some language errors depend on the context of the program's execution. Such errors

are called run-time errors or exceptions. Run-time errors arise after the interpreter's translation phase and during its execution phase.

The interpreter may issue an error for a syntactically correct statement like

```
x = y + 2
```

if the variable y has yet to be assigned; for example, if the statement appears
at line 12 and by that point y has not been assigned, we are informed:

```
>>> y = x + 2
Traceback (most recent call last):
    File "<pyshell#0>", line 1, in <module>
      y = x+ 2
NameError: name 'x' is not defined
```

Program 3.4 (division.py) shows an error that exists only in a limited case.

```
dividend = eval(input('Enter first numbers to divide:'))
divisor = eval(input('Enter second numbers to divide:'))
print(dividend, '/', divisor, "=", dividend/divisor)
```

In the above program the description dividend/divisor statement is confusing, so be careful to write proper statements. If the user enters, for example,
45 and 4, the program works nicely

```
Enter first numbers to divide: 45
Enter second numbers to divide: 4
45 / 4 = 11.25
```

When the user enters 32 and 0, the software declares an error and ends:

```
Enter first numbers to divide: 32
Enter second numbers to divide: 0
Traceback (most recent call last):
    File "C:\Users\Chanderbhan\Desktop\t.py", line 3, in
    <module>
      print(dividend, '/', divisor, "=", dividend/divisor)
    ZeroDivisionErro: division by zero
```

In algebra, division by zero is unknown, so division by zero is forbidden in Python. Program 3.5 is another way to derive it.

```
number = eval(input('Please enter a number to cut in
half: '))
# process the result
print(number/2)
```

Output

```
Please enter a number to cut in half: 150
75.0
```

and

```
Please enter a number to cut in half: 73.289
36.6445
```

So far, so good, but what if the user does not follow the on-screen instructions?

```
Please enter a number to cut in half: fun
Traceback (most recent call last):
    File "C:\Users\Chanderbhan\Desktop\t.py", line 1, in
    <module>
        number = eval(input('Please enter a number to cut
        in half: '))
    File "<string>", line 1, in <module>
    NameError: name 'fun' is not defined
```

or

```
Please enter a number to cut in half: 'fun'
Traceback (most recent call last):
    File "C:\Users\Chanderbhan\Desktop\t.py", line 3, in
    <module>
        print(number/2)
TypeError: unsupported operand type(s) for /: 'str' and
'int'
```

Since the programmer cannot predict what the user will provide as input, this program is doomed eventually. Fortunately, in this chapter we will examine techniques that allow programmers to avoid these kinds of problems.

The interpreter detects syntax errors immediately and the program never makes it out of the translation phase. Sometimes run-time errors do not reveal themselves immediately. The interpreter issues a run-time error only when it attempts to execute a statement with a problem. We will see how to write programs that optionally execute some statements only under certain conditions. If those conditions do not arise during testing, the faulty code is not executed. This means the error may lie undetected until a user stumbles upon it after the software is deployed. Run-time errors, therefore, are more troublesome than syntax errors.

3.4.3 Logic Errors

During the execution process, the translator can spot syntax errors and run-time errors. These include Python code infringements. These errors are best corrected by the programmer who is shown where the error precisely is in the source code.

Take into consideration the effect of replacing:

dividend/divisor;

with the following statement in Python program (division.py):

divisor/dividend;

The software is working, and the translator will not announce any errors except where a dividend value of null is reached. All the results it determines, though, are usually not accurate. Dividend = divisor is the only moment when the correct answer is written. The software produces an error, but it cannot be identified by the interpreter. A conceptual misunderstanding of this kind is recognized.

New programmers sometimes have issues with syntax and runtime errors early on because their vocabulary is limited. The error signals of the translator are also the best friend of the programmer. The amount of non-logical errors reduces or is trivially set and the amount of logic errors grows as programmers acquire knowledge of the language and written programs get more complex.

The translator, however, is unable to provide much input into the existence and position of logical errors. Consequently, reasoning errors appear to be the worst to identify and fix. Tools like debuggers also help find and repair logic errors, but such devices do not function automatically.

Unidentified flaws in runtime and reasoning defects are usually considered glitches in applications. The interpreter records errors in results only if the circumstances that show these errors are correct. For conceptual inconsistencies, the translator does not support them at all.

These glitches are the biggest source of developers' annoyance. The dissatisfaction is often caused by the fact that errors occur in complex programs only in certain situations that are difficult to replicate during tests precisely. Such annoyance is realized as the systems get more complex.

The positive news is the systematic implementation of effective design practices to the programming environment will help to raise the number of variations of found logical errors. Check properly where the logic is violated and correct. The bad news is that logical errors are inevitable because of the creation of software in an essentially human conceptual quest. The implementation and eventual analysis and removal of logical errors are an important part of the programming method.

3.5 Examples of Arithmetic

Suppose we wish to convert temperature from degrees Fahrenheit to degrees Celsius. The following formula provides the necessary mathematics:

$$°C = \frac{5}{9} \times \left(°F - 32\right)$$

The conversion program (CelsiustoFahrenheit.py) is:

```
fahrenheit = float(input("Enter temperature in
fahrenheit:"))
celsius = (fahrenheit - 32) * 5/9
print('%.2f fahrenheit is: %0.2f Celsius' %(fahrenheit,
celsius))
```

Output

```
Enter temperature in fahrenheit: 98
98.00 Fahrenheit is: 36.67 Celsius
```

Program (CelsiustoFahrenheit.py) shows a program to calculate Celsius and Fahrenheit temperature conversion. Several instances show how the software works:

```
seconds = eval(input("Please enter the number of
seconds:"))
# First, compute the number of hours in the given number
of seconds
# Note: integer division with possible truncation
hours = seconds // 3600 # 3600 seconds = 1 hours
# Compute the remaining seconds after the hours are
accounted for
seconds = seconds % 3600
# Next, compute the number of minutes in the remaining
number of seconds
minutes = seconds // 60 # 60 seconds = 1 minute
# Compute the remaining seconds after the minutes are
accounted for
seconds = seconds % 60
# Report the results
print(hours, "hr,", minutes, "min,", seconds, "sec")
```

Output

```
Please enter the number of seconds: 10000
2 hr, 46 min, 40 sec
```

If the user enters 10000, the program prints 2 hr, 46 min, 40 sec. Notice the assignments to the seconds variable, such as

```
seconds = seconds % 3600
```

The right side of the assignment operator (=) is first evaluated. The remainder of seconds divided by 3,600 is assigned back to seconds. This statement can alter the value of seconds if the current value of seconds is greater than 3,600. A similar statement that occurs frequently in programs is one like

```
x = x + 1
```

This statement increments the variable x to make it one bigger. A statement like this one provides further evidence that the Python assignment operator does not mean mathematical equality. The following statement from mathematics

```
x = x + 1
```

surely is never true; a number cannot be equal to one more than itself. If that were the case, I would deposit one dollar in the bank and then insist that I really had two dollars in the bank, since a number is equal to one more than itself. If two dollars were to become three and then four, etc., I would soon be rich. In Python, however, this statement simply means "add one to x's current value and update x with the result."

A variation on program (timeconv.py), is program (enhancedtimeconv.py) which performs the same logic to compute time components (hours, minutes and seconds), but it uses simpler arithmetic to produce a slightly different output: instead of printing 11,045 seconds as 3 hr, 4 min, 5 sec, program (enhancedtimeconv.py) displays it as 3:04:05. It is trivial to modify program (timeconv.py) so that it prints 3:4:5, but program (enhancedtimeconv.py) includes some extra arithmetic to put leading zeroes in front of single-digit values for minutes and seconds as is done on digital clock displays.

```
seconds = eval(input("Please enter the number of
seconds:"))
# First, compute the number of hours in the given number
of seconds
# Note: integer division with possible truncation
hours = seconds // 3600 # 3600 seconds = 1 hours
# Compute the remaining seconds after the hours are
accounted for
seconds = seconds % 3600
# Next, compute the number of minutes in the remaining
number of seconds
minutes = seconds // 60 # 60 seconds = 1 minute
# Compute the remaining seconds after the minutes are
accounted for
seconds = seconds % 60
# Report the results
print(hours, ":", sep="", end="")
# Compute tens digit of minutes
tens = minutes // 10
# Compute ones digit of minutes
ones = minutes % 10
print(tens, ones, ":", sep="", end="")
# Compute tens digit of seconds
tens = seconds // 10
# Compute ones digit of seconds
ones = seconds % 10
print(tens, ones, sep ="")
```

Output

```
Please enter the number of seconds: 10000
2: 46: 40
```

Program (converttime.py) uses the fact that if x is a one- or two-digit number, x % 10 is the tens digit of x. If x % 10 is zero, x is necessarily a one-digit number. If x for 10 is empty, x is a one-digit number.

3.6 Algorithms

An algorithm is a finite series of steps, where each step takes a certain time, solves a topic or a meaning is calculated. One example of an algorithm is a programming machine, which computes a result. A series of measures apply to both of these instances. In the case of lasagna, it is important to cook the pasta in boiling water before it is put onto the baked filling. The raw nutrients and all the other components in the bowl will be baked and then discarded since the baked nutrients have been cooked separately in the boiling water. Additionally, the sequence of actions in a computer program is also significant. While this is clear, consider the following logical statement:

- The relation between Celsius and Fahrenheit can be described as

$$°C = \frac{5}{9} \times \left(°F - 32 \right)$$

- Provided with a Fahrenheit temperature, the correct temperature may be measured in degrees Celsius.

Program 3.9 (error.py)

```
degreesF, degreesC = 0, 0
# Define the relationship between F and C
degreesC = 5/9*(degreesF - 32)
# Prompt user for degrees F
degreesF = eval(input('Enter the temperature in
degrees F: '))
# Report the result
print(degreesF, "degrees F =', degreesC, 'degrees C'")
```

Output

```
Enter the temperature in degrees F: 96
96 degrees F = ', degreesC, 'degreesC'
```

Irrespective of the given data, the above definition in English is true. The formula is diligently applied. The problem lies simply in statement ordering. The statement

```
degrees C = 5/9*(degrees F - 32);
```

is an assignment argument, and not a definition of a relationship that exists throughout the program. At the point of the assignment, degreesF has the value of zero. The variable degreesC is assigned *before* degreesF's value is received from the user.

Consider x and y in certain programs which are two variables. Why will the values of all variables be exchanged? We want x to have the original value of y and the original value of y to have x. The code can seem reasonable:

```
x = y
y = x
```

The concern with this code segment is that x and y have the same value since the first line has been implemented (the initial value is y). The second function is superfluous and affects neither of the x or y values. A third variable is needed until the solution reassigns the original value of one of the variables. The best coding for modifying values is

```
temp = x
x = y
y = temp
```

We will render the swap simpler with the use of tuple assignments (see Chapter 2):

```
x, y = y, x
```

Such limited details underline the accuracy of algorithms. Informal ideas of how to address a problem can be helpful in the early stages of software design, but a thorough explanation of the approach is needed by the coded system.

The algorithms we have seen so far were clear. Statement 1, then Statement 2, etc., before all system statements have been implemented, introduce language constructs that require certain statements to be optionally and repetitively implemented. These frameworks help one to create programs that do a lot of interesting stuff, yet also involves sophisticated algorithms. Bear in mind the reality that 99 percent of a complex algorithm is incorrect. The architecture and implementation of an algorithm may be halted by inattention.

3.7 Conclusion

This chapter has described various types of complicated arithmetic expressions involving many operators and operands, the rules pertaining to mixed arithmetic have been applied on an operator-by-operator basis and used on various types of operators in Python, such as arithmetic, logical, assignment, bitwise, membership, identity and comparison operators. Python supports a lot of operators to help us use common arithmetic, comparison, assignment, binary, or logical operations. The operators can be used by defining special methods. Assignment operators are crucial elements in Python programming. They are quite useful in situations that involve breaking bigger tasks into smaller units. The application of these operators is evident in large-scale programs where the program code needs to be optimized to the greatest possible extent.

4

Branch Control Structure

4.1 Boolean Expressions

Arithmetic expressions evaluate numeric values; a Boolean expression, sometimes called a predicate, may have only one of two possible values: false or true. The term Boolean comes from the name of the British mathematician George Boole. A branch of discrete mathematics called Boolean algebra is dedicated to the study of the properties and the manipulation of logical expressions. While on the surface Boolean expressions may appear very limited compared to numeric expressions, they are essential for building more interesting and useful programs [18].

The simplest Boolean expressions in Python are `True` and `False`. In a Python interactive shell we see:

```
>>> True
True
>>> False
False
>>> type(True)
<class 'bool'>
>>> type(False)
<class 'bool'>
```

We see that `bool` is the name of the class representing Python's Boolean expressions. Program 4.1 (boolvars.py) above is a simple program that shows how Boolean variables can be used.

DOI: 10.1201/9781003202035-4

TABLE 4.1

Python Relational Operators

Expression	Meaning
$x == y$	True if $x = y$ (mathematical equality, not assignment); otherwise, false
$x < y$	True if $x < y$; otherwise, false
$x <= y$	True if $x \leq y$; otherwise, false
$x > y$	True if $x > y$; otherwise, false
$x >= y$	True if $x \geq y$; otherwise, false
$x != y$	True if $x 6 = y$; otherwise, false

TABLE 4.2

Examples of Simple Relational Expressions

Expression	Value
10 < 20	True
10 >= 20	False
x < 100	True if x is less than 100; otherwise, false
x != y	True unless x and y are equal

```
# Assign some Boolean variables
a = True
b = False
print('a =', a, ' b =', b)
# Reassign a
a = False;
print('a =', a, ' b =', b)
```

Output

```
a = True b = False
a = False b = False
```

4.2 Additional Boolean Statements

We have seen that the simplest Boolean expressions are false and true – the Python Boolean literals. A Boolean variable is also a Boolean expression. An expression comparing numeric expressions for equality or inequality is also a Boolean expression. These comparisons are done using relational operators. Table 4.1 lists the relational operators available in Python.

Table 4.2 shows some simple Boolean expressions with their associated values. An expression like 10 < 20 is legal but of little use, since 10 < 20 is always true; the expression True is equivalent, simpler and less likely to confuse human readers. Since variables can change their values during a program's execution, Boolean expressions are most useful when their truth values depend on the values of one or more variables.

The relational operators are binary operators and are all left associative. They all have a lower precedence than any of the arithmetic operators; therefore, the expression

```
x + 2 < y / 10
```

is evaluated as if parentheses were placed as:

```
(x + 2) < (y / 10)
```

4.3 The Simple If Statement

The Boolean expressions described in Section 4.2 at first may seem arcane and of little use in practical programs. In reality, Boolean expressions are essential for a program to be able to adapt its behavior at run time. Most truly useful and practical programs would be impossible without the availability of Boolean expressions.

The execution errors mentioned in Section 3.4 arise from logic errors. One way that Program 3.4 (dividedanger.py) can fail is when the user enters a zero for the divisor. Fortunately, programmers can take steps to ensure that division by zero does not occur. Program 4.2 (betterdivision.py) shows how it might be done [19].

```
dividend, divisor = eval(input('Enter two numbers to
divide: '))
# If possible, divide them and report the result
if divisor != 0:
print(dividend, '/', divisor, "=", dividend/divisor)
```

Output
```
Please enter two numbers to divide: 16, 8
16/8=2.0
```

The program prints nothing after the user enters the values.

The last non-indented line in Program 4.2 (betterdivision.py) begins with the reserved word `if`. The `if` statement allows code to be optionally executed. In this case, the printing statement is executed only if the variable divisor's value is not zero.

The Boolean expression

```
divisor != 0
```

determines if the statement in the indented block that follows is executed. If the divisor is not zero, the message is printed; otherwise, nothing is displayed.

Figure 4.1 shows how program execution flows through the `if` statement of Program 4.2 (betterdivision.py).

The basic form of the declaration is:

- The reserved word `if` begins an `if` statement.
- The condition is a Boolean expression that determines whether or not the body will be executed. A colon (:) must follow the condition.
- The block is a block of one or more statements to be executed if the condition is true. Recall that the statements within a block must all be indented by the same number of spaces from the left. The block within an `if` must be indented by more spaces than the line that begins the `if` statement. The block technically is part of the `if` statement. This part of the `if` statement is sometimes called the body of the `if` [19].

Python requires the block to be indented. If the block contains just one statement, some programmers will place it on the same line as the `if`; for example, in the following `if` statement that optionally assigns y,

```
if x < 10: y = x
```

could be written

```
if x < 10: y = x
```

but may not be written as

```
if x < 10: y = x
```

because the lack of indentation hides the fact that the assignment statement is optionally executed. Indentation is how Python determines which statements make up a block.

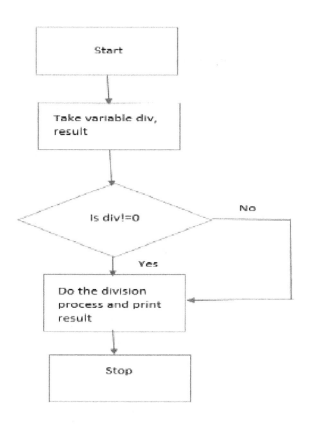

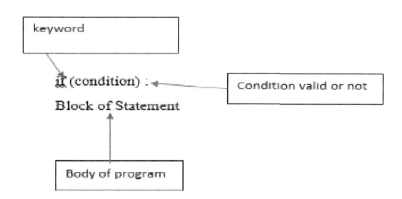

FIGURE 4.1
If Statement flowchart.

It is important not to mix spaces and tabs when indenting statements in a block. In many editors you cannot visually distinguish between a tab and a sequence of spaces. The number of spaces equivalent to the spacing of a tab differs from one editor to another. Most programming editors have a setting to substitute a specified number of spaces for each tab character. For Python development you should use this feature. It is best to eliminate all tabs within your Python source code.

How many spaces should you indent? Python requires at least one; some programmers consistently use two; four is the most popular number; but some prefer a more dramatic display and use eight. A four-space indentation for a block is the recommended Python style. This text uses the recommended four spaces to set off each enclosed block. In most programming editors you can set the tab key to insert spaces automatically so you need not count the spaces as you type.

The `if` block may contain multiple statements to be optionally executed. Program 4.3 (alternatedivision.py) optionally executes two statements depending on the input values provided by the user.

```
dividend = eval(input("Enter first number to divide:"))
divisor = eval(input("Enter second number to divide:"))
if divisor !=0:
    quotient = dividend/divisor
    print(dividend, '/', divisor, "=", quotient)
print("Program finished")
```

Output

```
Enter first numbers to divide: 55
Enter second numbers to divide: 5
55 / 5 = 11.0
Program finished
```

The assignment statement and first printing statement are both a part of the block of the `if`. Given the truth value of the Boolean expression `divisor != 0` during a particular program run, either both statements will be executed or neither statement will be executed. The last statement is not indented, so it is not part of the `if` block. The program always prints "Program finished", regardless of the user's input.

Remember when checking for equality, as in

```
if x == 11:
    print('eleven')
```

always to use the relational equality operator (==), and not the assignment operator (=).

As a convenience to programmers, Python's notion of true and false extends beyond what we ordinarily would consider Boolean expressions. The statement

```
if 1:
    print('one')
```

always prints "one", while the statement

```
if 0:
    print('zero')
```

never prints anything. Python considers the integer value zero to be false and any other integer value to be true. Similarly, the floating-point value 0.0 is false, but any other floating-point value is true. The empty string ('' or "") is considered false, and any non-empty string is interpreted as true. Any Python expression can serve as the condition for an if statement. In later chapters we will explore additional kinds of expressions and see how they relate to Boolean conditions [20].

Program 4.4 (leadingzeros.py) requests an integer value from the user. The program then displays the number using exactly four digits. The program prepends leading zeros where necessary to ensure all four digits are occupied. The program treats numbers less than zero as zero and numbers greater than 9,999 as 9999.

```
num = eval(input("Please enter an integer in the range
0...9999"))
if num < 0:
    num = 0
if num > 9999:
    num = 9999
    print(end="[")
        digit = num//1000
    print(digit, end="")
    num %=1000
    digit = num//100
    print(digit, end="")
    num %=100
    digit = num//10
    print(digit, end="")
    num %=10
    print(num, end="")
```

Output

```
Please enter an integer in the range 0...9999: 100000
[9999
```

Program 4.4 (rangeinteger.py) allows you to provide an integer value. The system then displays a number of four digits exactly by including leading nulls. Numbers less than zero and numbers more than 9,999 are considered as 9999 by the program itself.

```
print ("]") # Print right brace
```

The two lists in Program 4.4 (rangeinteger.py) force the number to be within range if the statements are at first. certain mathematical claims containing portions of the number to be shown. Remember the declaration

```
num %= 10
```

is short for

```
num = num % 10
```

4.4 If/Else Control Statements

One undesirable aspect of Program 4.2 (division.py) is if the user enters a zero divisor, nothing is printed. It may be better to provide some feedback to the user to indicate that the divisor provided cannot be used [21]. The if statement has an optional else clause that is executed only if the Boolean condition is false. Program 4.5 (feedback.py) uses the if/else statement to provide the desired effect.

```
divisor = eval (input ('Please enter two numbers to
divide:'))
dividend = 0
if divisor !=0:
print (dividend, '/', divisor, "=", dividend/divisor)
else:
print ("Division by zero is not allowed")
```

Output

```
Please enter two integers to divide: 32
0/32=0.0
```

The else clause contains an alternate block that is executed if the condition is false. The program's flow of execution is shown in Figure 4.2.

Program 4.5 (feedback.py) avoids the division by using a zero run-time error that causes the program to terminate prematurely, though it still alerts the user that there is a problem. Another application may handle the situation in a different way; for example, it may substitute some default value for the divisor instead of zero.

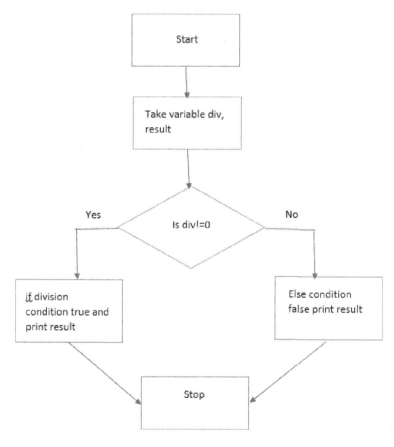

FIGURE 4.2
If/else flowchart.

The general form of an if/else statement is

```
if condition :
      if-Statement
else :
      else-Statement
```

- The reserved word if begins the if/else statement.
- The condition is a Boolean expression that determines whether or not the if block or the else block are executed. A colon (:) must follow the condition.
- The if block is a block of one or more statements to be executed if the condition is true. As with all blocks, it must be indented by more spaces than the if line. This part of the if statement is sometimes called the body of the if.
- The reserved word else begins the second part of the if/else statement. A colon (:) must follow the else.
- The else block is a block of one or more statements to be executed if the condition is false. It must be indented more spaces than the else line. This part of the if/else statement is sometimes called the body of the else.

The else block, like the if block, consists of one or more statements indented to the same level.

The equality operator (==) checks for exact equality. This can be a problem with floating-point numbers, since they are inherently imprecise. Program 4.6 (samedifferent.py) demonstrates the perils of using the equality operator with floating-point numbers.

```
d1 = 1.11 - 1.10 d2 = 2.11 - 2.10
print('d1 =', d1, ' d2 =', d2)
if d1 == d2:
print('Same')
else:
print('Different')
```

In mathematics, we expect the following equality to hold:

```
1.11 = 1.10 = 0.01 = 2.11 - 2.10
```

The output of the first print statement reminds us of the imprecision of floating-point numbers:

```
d1 = 0.010000000000000009
d2 = 0.009999999999999787
```

Since the expression

```
d1 == d2
```

checks for exact equality, the program reports that d1 and d2 are different. Later we will see how to determine if two floating-point numbers are "close enough" to be considered equal.

4.5 Compound Boolean Expressions

Simple Boolean expressions, each involving one relational operator, can be combined into more complex Boolean expressions using the logical operators and, or and not. A combination of two or more Boolean expressions using logical operators is called a compound Boolean expression [22].

To introduce compound Boolean expressions, consider a computer science degree that requires, among other computing courses, Operating Systems and Programming Languages. If we isolate those two courses, we can say a student must successfully complete both Operating Systems and Programming Languages to qualify for the degree. A student that passes Operating Systems but not Programming Languages will not have met the requirements. Similarly, Programming Languages without Operating Systems is insufficient, and a student completing neither Operating Systems nor Programming Languages surely does not qualify.

Logical AND works in exactly the same way. If e1 and e2 are two Boolean expressions, e1 and e2 is true only if e1 and e2 are both true; if either one is false or both are false, the compound expression is false.

To illustrate logical OR, consider two mathematics courses, Differential Equations and Linear Algebra. A computer science degree requires one of those two courses. A student who successfully completes Differential Equations but does not take Linear Algebra meets the requirement. Similarly, a student may take Linear Algebra but not Differential Equations. A student that takes neither Differential Equations nor Linear Algebra certainly has not met the requirement. It is important to note the a student may elect to take both Differential Equations and Linear Algebra (perhaps on the way to a mathematics minor), but the requirement is no less fulfilled.

TABLE 4.3

Operators of Logical Boolean Expressions are e_1 and e_2

e1	e2	e1 and e2	e1 or e2	not e1
False	False	False	False	True
False	True	False	True	True
True	False	False	True	False
True	True	True	True	False

Logical OR works in a similar fashion. Given our Boolean expressions e1 and e2, the compound expression e1or e2 is false only if e1 and e2 are both false; if either one is true or both are true, the compound expression is true. Note that logical OR is an inclusive or, not an exclusive or. In informal conversation we often imply exclusive or, as in a statement like "Would you like cake or ice cream for dessert?" The implication is one or the other, not both. In computer programming the or is inclusive; if both subexpressions in an or expression are true, the or expression is true [23].

Logical NOT simply reverses the truth value of the expression to which it is applied. If e is a true Boolean expression, not e is false; if e is false, not e is true.

Table 4.3 is called a truth table. It shows all the combinations of truth values for two Boolean expressions and the values of compound Boolean expressions built from applying the and, or and not Python logical operators.

Both and and or are binary operators; that is, they require two operands, both of which must be Boolean expressions. The not operator is a unary operator (see Section 3.1); it requires a single Boolean operand immediately to its right.

Operator not has higher precedence than both and and or, and has higher precedence than or; and and or are left associative; not is right associative; and and or have lower precedence than any other binary operator except assignment. This means the expression

```
x <= y and x <= z
```

is evaluated as

```
(x <= y) and (x <= z)
```

Some programmers prefer to use the parentheses as shown here even though they are not required. The parentheses improve the readability of complex expressions, and the compiled code is no less efficient.

Python allows an expression like

```
x <= y and y <= z
```

which means x ≤ y ≤ z is expressed more naturally:

```
x <= y <= z
```

Similarly, Python allows a programmer to test the equivalence of three variables as

```
if x == y == z:
    print('They are all the same')
```

The following section of code assigns the indicated values to a bool:

```
x = 10 y = 20
b = (x == 10)                      #    assigns True to b
b = (x != 10)                      #    assigns False, to b
b = (x == 10 and y == 20)          #    assigns True to b
b = (x != 10 and y == 20)          #    assigns False to b
b = (x == 10 and y != 20)          #    assigns False to b
b = (x != 10 and y != 20)          #    assigns False to b
b = (x == 10 or y == 20)           #    assigns True to b
b = (x != 10 or y == 20)           #    assigns True to b
b = (x == 10 or y != 20)           #    assigns True to b
b = (x != 10 or y != 20)           #    assigns False to b
```

Convince yourself that the following expressions are equivalent:

```
x != y
```

and

```
!(x == y)
```

and

```
x < y or x > y
```

In the expression e1 and e2 both subexpressions e1 and e2 must be true for the overall expression to be true. Since the and operator evaluates from left to right, this means that if e1 is false, there is no need to evaluate e2. If e1 is false, no value of e2 can make the expression e1 and e2 true. The and operator first tests the expression to its left. If it finds the expression to be false, it does not bother to check the right expression. This approach is called short-circuit evaluation. In a similar fashion, in the expression e1 or e2, if e1 is true, then it does not matter what value e2 has – an or expression is true unless both subexpressions are false. The or operator uses short-circuit evaluation also.

Why is short-circuit evaluation important? Two situations show why:

- The order of the subexpressions can affect performance. When a program is running, complex expressions require more time for the computer to evaluate than simpler expressions. We classify an expression that takes a relatively long time to evaluate as an expensive expression. If a compound Boolean expression is made up of an expensive Boolean subexpression and a less expensive Boolean subexpression, and the order of evaluation of the two expressions does not affect the behavior of the program, then place the more expensive Boolean expression second. If the first subexpression is False and the operator and is being used, then the expensive second subexpression is not evaluated; if the first subexpression is True and the or operator is being used, then, again, the expensive second subexpression is avoided.

- Subexpressions can be ordered to prevent run-time errors. This is especially true when one of the subexpressions depends on the other in some way. Consider the following expression:

```
(x != 0) and (z/x > 1)
```

Here, if x is zero, the division by zero is avoided. If the subexpressions were switched, a run-time error would result if x is zero.

Suppose you wish to print the word "OK" if a variable x is 1, 2 or 3. An informal translation from English might yield:

```
if x == 1 or 2 or 3: print("OK")
```

Unfortunately, x's value is irrelevant; the code always prints the word "OK" regardless of the value of x. Since the == operator has lower precedence than ||, the expression

```
x == 1 or 2 or 3
```

is interpreted as

```
(x == 1) or 2 or 3
```

The expression x == 1 is either true or false, but integer 2 is always inter-preted as true, and integer 3 is interpreted as true as well. If x is known to be an integer and not a floating-point number, the expression

```
1 <= x <= 3
```

also would work.

The correct statement would be

```
if x == 1 or x == 2 or x == 3: print("OK")
```

The revised Boolean expression is more verbose and less similar to the English rendition, but it is the correct formulation for Python.

4.6 Nested If/Else Conditional Statements

The statements in a block of the `if` or the `else` may be any Python state-ments, including other if/else statements. These nested `if` statements can be used to develop an arbitrarily complex control flow logic. Consider Program 4.7 (matchrange.py) that determines if a number is between 0 and 10, inclu-sive [24].

```
value = eval(input("enter an integer value In the
range 0...10: ")
if value >= 0: # First check
    if value <= 10:# Second check print ("In range")
print("Done")
```

- The first condition is checked. If the value is less than zero, the second condition is not evaluated and the statement following the outer `if` is executed. The statement after the outer `if` simply prints "Done".

- If the first condition finds a value to be greater than or equal to zero, the second condition is then checked. If the second condition is met, the "In range" message is displayed; otherwise, it is not. Regardless, the program eventually prints the "Done" message.

We say that the second `if` is nested within the first `if`. We call the first `if` the outer `if` and the second `if` the inner `if`. Both conditions of this nested `if` construct must be met for the "In range" message to be printed. Said

Python for Beginners

another way, the first condition and the second condition must be met for the "In range" message to be printed. From this perspective, the program can be rewritten to behave the same way with only one if statement, as Program 4.8 (currentmatchrange.py) shows.

```
value = eval(input("Enter an integer value in the range
0...10: ")
    if value >= 0 and value <= 10: # Only one, more
    complicated check
    print("In range")
    print("Done")
```

Here all conditions are verified concurrently using the operator and the variable. The reasoning, utilizing just an if statement becomes an easier Boolean expression at the expense of a much more complex one. The second version is superior since simplistic thought generally serves a constructive function. In Program 4.8 (newcheckrange.py) check the conditions:

```
value >= 0 and value <=10
    and be expressed more compactly as
        0 <= value <=10
```

Often, as in this program, the rationale is not condensed. It would be impossible to rewrite Program 4.10 (enhancematchrange.py) with a single if statement.

```
value = eval(input("enter an integer value in the range
0....10:"))
if value >= 0:
   if value <=10:
   print(value, "is in range")
else:
print(value, "is too large")
else:
print(value, "is too small")
print ("Done")
```

Output

```
enter an integer value in the range
0....10:9
9 is in range
```

This program includes a more detailed note rather than a clear approval notification. The variable value precisely prints one of three texts. A single declaration if or if/else is chosen from more than two routes.

A simple troubleshooter is provided in program 4.9 (choiceprogram.py) and could be used by a computer technician to treat a troubled device.

```
print("Help! My computer doesn't work")
print("Does the computer make any sounds (fans, etc.)")
  choice = input("or show any lights? (y/n):")
if choice =='n':
  choice = input("Is it plugged in? (y/n):")
if choice =='n':
  print("plug it in. If the problem persists, ")
print("please run this program again.")
else:
choice = input("Is the switch in the \"on\" position?
(y/n):")
if choice == 'n':
print("Turn it on. If the problem persists,")
print("please run this program again.")
else:
choice = input("Does tje computer have a fuse? (y/n):")
  if choice == 'n':
choice = input("Is the outlet's OK? (y/n)")
if choice == 'n':
print("Check the outlet's circuit")
print("btraker or fuse. Move to a")
print("new outlet, if necessary.")
print("If the problem persists")
print("Please run this program again.")
else:
print("Please consult a service technician.")
'''else:
print("Check the fuse. Replace if")
print("necessary. If the problem")
print("persists; then")
print("Please run this program again.")
else:
print("Please consult a service technician.")'''
```

Output

```
Help! My computer doesn't work
Does the computer make any sounds (fans, etc.)
or show any lights (y/n):y
```

```
Is ther switch in the "on" position? (y/n):y
Does the computer have a fuse? (y/n):n
Is the outlet's OK? (y/n):n
Check the outlet's circuit
btraker or fuse. Move to a
new outlet, if necessary.
If the problem persists
Please run this program again.
```

Really basic software attempts to determine why a machine doesn't function. The potential for enhancement is unlimited, but this version deals only with power issues that have simple fixes. Notice that if the computer has power (fan or disk drive makes sounds or lights are visible), the program indicates that help should be sought elsewhere! The decision tree capturing the basic logic of the program is shown in Figure 4.3. The steps performed are:

- Is it plugged in? This simple fix is sometimes overlooked.
- Is the switch in the "on" position? This is another simple fix.
- If applicable, is the fuse blown? Some computer systems have a user-serviceable fuse that can blow out during a power surge. (Most newer computers have power supplies that can handle surges and have no user-serviceable fuses.)
- Is there power at the receptacle? Perhaps the outlet's circuit breaker or fuse has a problem.

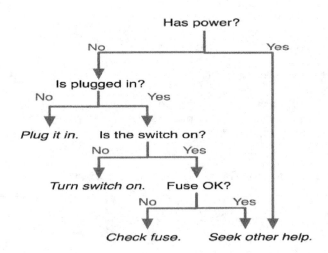

FIGURE 4.3
Decision tree for operating device troubleshooting.

The easiest checks are made first. Progressively more difficult checks are introduced as the program runs. Based on your experience with trouble-shooting computers that do not run properly, you may be able to think of many enhancements to this simple program.

Note the various blocks of code and how the blocks are indented. Visually programmers can quickly determine the logical structure of a program by the arrangement of the blocks.

4.7 Multipile Decision-Making Statements

A simple if/else statement can select from between two execution paths. Program 4.8 (currentmatchrange.py) showed how to select from among three options. What if exactly one of many actions should be taken? Nested if/else statements are required, and the form of these nested if/else statements is shown in Program 4.11 (digittoword.py) [25].

```python
value = eval(input("Enter an integer in the range 0...5:"))
if value < 0:
 print("Too small")
elif value == 0:
 print("Zero")
elif value == 1:
 print("one")
elif value == 2:
 print("two")
elif value == 3:
 print("three")
elif value == 4:
 print("four")
elif value == 5:
 print("five")
else:
 print("Too large")
 print("Done")
```

Output

```
Enter an integer in the range 0...5:5
five
```

Observe the following about Program 4.11 (digittoword.py):

- It prints exactly one of eight messages depending on the user's input.
- Notice that each `if` block contains a single printing statement and each `else` block, except the last one, contains an `if` statement. The control logic forces the program execution to check each condition in turn. The first condition that matches wins, and its corresponding `if` body will be executed. If none of the conditions are true, the program prints the last `else`'s "Too large" message.

As a consequence of the required formatting of Program 4.11 (digittoword. py), the mass of text drifts to the right as more conditions are checked. Python provides a multi-way conditional construct called if/elif/else that permits a more manageable textual structure for programs that must check many conditions. Program 4.12 (restyleddigittoword.py) uses the if/elif/else statement to avoid the rightward code drift.

The word `elif` is a contraction of `else` and `if`; if you read `elif` as `else if`, you can see how the code fragment

```
else: if value == 2:
        print("two")
```
in Program 4.11 (digittoword.py) can be transformed into

```
elif value == 2:
        print("two")
```

in Program 4.12 (restyleddigittoword.py).

The if/elif/else statement is valuable for selecting exactly one block of code to execute from several different options. The `if` part of an if/elif/else statement is mandatory. The `else` part is optional. After the `if` part and before the `else` part (if present) you may use as many `elif` blocks as necessary.

Program 4.13 (datetransformer.py) uses an if/elif/else statement to transform a numeric date in month/day format to an expanded US English form and an international Spanish form; for example, 2/14 would be converted to February 14 and 14 febrero.

```
month = eval(input("Please enter the month as a number
(1-12):"))
day = eval(input("Please enter the day of the month:"))
if month == 1: print("January", end='')
elif month == 2: print("February", end='')
elif month == 3: print("March", end='')
elif month == 4: print("April", end='')
elif month == 5: print("May", end='')
```

```
elif month == 6: print("June", end='')
elif month == 7: print("July", end='')
elif month == 8: print("August", end='')
elif month == 9: print("September", end='')
elif month == 10: print("October", end='')
elif month -- 11: print("November", end='')
else:
    print("December", end='')

print(day, 'or', day, end='')
if month == 1: print("enero")
elif month == 2: print("febrero")
elif month == 3: print("Marzo")
elif month == 4: print("abril")
elif month == 5: print("mayo")
elif month == 6: print("Junio")
elif month == 7: print("Julio")
elif month == 8: print("agosto")
elif month == 9: print("Septiembre")
elif month == 10: print("Octubre")
elif month == 11: print("Noviembre", end='')
else:
    print("dember")
```

Output

```
Please enter the month as a number (1-12): 5
Please enter the day of the month: 5
May 20 or 20 mayo
```

4.8 Expressions of Decision-Making Conditional Statements

Consider the following code fragment:

```
if a != b:
    c = d else:
    c = e
```

Here variable c is assigned one of two possible values. As purely a syntactical convenience, Python provides an alternative to the if/else construct called a conditional expression. A conditional expression evaluates one of two values depending on a Boolean condition. The above code can be rewritten as

```
c = d if a != b else e
```

The general form of the conditional expression is

```
expression1 if condition else expression2
```

where

- `expression1` is the overall value of the conditional expression if the condition is true;
- `condition` is a normal Boolean expression that might appear in an `if` statement;
- `expression2` is the overall value of the conditional expression if the condition is false.

Program 4.14 (safedivide.py) uses our familiar if/else statement to check for division by zero

```
dividend = eval(input("Enter dividend, divisor:"))
divisor = eval(input("Enter dividend, divisor:"))
if divisor != 0:
  print(dividend/divisor)
else:
  print("Error, cannot divide by zero")

dividend = eval(input("Enter dividend, divisor:"))
divisor = eval(input("Enter dividend, divisor:"))
msg = dividend/divisor
if divisor != 0:
  print(msg)
```

Output

```
Enter dividend, divisor: 56
Enter dividend, divisor: 5
11.2
Enter dividend, divisor: 56
Enter dividend, divisor: 5
11.2
```

Note that the type of msg variable depends on the expression given in program (safedivideconditional.py), the msg value ('Error, cannot The splitting point (dividend/divisor), or the series (dividend/dividend).

In the case of mathematics, another example is the absolute value of the number:

$$|n| = \begin{cases} n, & \text{when } n \geq 0 \\ -n, & \text{when } n < 0 \end{cases}$$

The real value of a variable or empty number is the same as the number; the total value of the other extra value is the negative addition. The Python expression displays the vector n as absolute:

```
-n if n < 0 else n
```

The word is not an argument. Program 4.14 (absvalue.py) is simple software explaining the conditional language used in a sentence.

```
# Acquire a number from the user and print its absolute
value.
n = eval(input("Entex a number: "))
    print ('|' , n, ' | = ' , (-n if n < 0 else n), sep=")
```

Output

```
Enter a number: -34
|-34| = 34
and
Enter a number: 0
|0| = 0
and
Enter a number: 100
|100| = 100
```

Some claim that the word "conditional" is just not as understandable as a standard if/else declaration. However, because of its very unique existence, it is used sparingly. Standard if/else blocks can include several claims, but the content is restricted in the conditional expression to specific words [26].

4.9 Errors in Decision-Making Statements

Carefully consider each compound conditional used, such as

```
value > 0 and value <= 10
```

found in Program 4.8 (newcheckrange.py). Confusing logical and and logical or is a common programming error. Consider the Boolean expression

```
x > 0 or x <= 10
```

What values of x make the expression true, and what values of x make the expression false? The expression is always true, no matter what value is assigned to the variable x. A Boolean expression that is always true is known as a tautology. Think about it. If x is a number, what value could the variable x assume that would make this Boolean expression false? Regardless of its value, one or both of the subexpressions will be true, so the compound logical or expression is always true. This particular or expression is just a complicated way of expressing the value True.

Another common error is contriving compound Boolean expressions that are always false, known as contradictions. Suppose you wish to exclude values from a given range; for example, reject values in the range 0…10 and accept all other numbers. Is the Boolean expression in the following code fragment up to the task?

```
# All but 0, 1, 2, ..., 10 if value < 0 and value
> 10: print(value)
```

A closer look at the condition reveals it can never be true. What number can be both less than zero and greater than ten at the same time? None can, of course, so the expression is a contradiction and a complicated way of expressing False. To correct this code fragment, replace the and operator with or [27].

4.10 Conclusion

In this chapter, control structures have been seen to allow control of the flow of program execution. There are four fundamentals of control structure. The control structure selection is used to select one of many possible paths of execution in a program depending on the given condition. Each condition is expressed in Python by a bool expression. Selection can be expressed by the if statement or if-else statement. The if statement decides whether or not a code block is to be executed. It decides whether or not a code block is to be executed. The if-else statement selects between two possible code blocks to be executed.

5

Iterative Control Structure

Looping is a program or script procedure that performs repetitive (or iterative) tasks. The loop is a sequence of instructions (a block) that is executed repeatedly while or until some condition is met or satisfied. Each repetition is called an iteration. All programming and scripting languages provide loop structures, and, like other languages, Python has several looping constructs [28].

Algorithm design is frequently based on iteration and conditional execution.

5.1 The While Loop

The output of Program 5.1 (counting.py) counts up to five.

```
print (5)
print (4)
print (3)
print (2)
print (1)
```

Output

```
5
4
3
2
1
```

How would you write the code to count to 10,000? Would you copy, paste, and modify 10,000 printing statements? You could, but that would be impractical! Counting is such a common activity, and computers routinely count up to very large values, that there must be a better way. What we would really like to do is print the value of a variable (call it count), then increment the

DOI: 10.1201/9781003202035-5

variable (count += 1), repeating this process until the variable is large enough (count == 5 or maybe count == 10000). Python has two different keywords, while and for, that enable iteration [29].

Program 5.2 (loopcounting.py) uses the keyword while to count up to five:

```
count = 1
while count <= 5:
  print (count)
  count += 1
```

Output

```
1
2
3
4
5
```

The while keyword in Program 5.2 (loopcounting.py) causes the variable 'count' to be repeatedly printed then incremented until its value reaches five. The block of statements

```
print (count)
count += 1
```

is executed five times. After each print of the variable count, the program increments it by one. Eventually (after five iterations), the condition count <= 5 becomes false, and the block is no longer executed.

Unlike the approach taken in Listing 5.1 (counttofive.py), it is trivial to modify Listing 5.2 (iterativecounttofive.py) to count up to 10,000—just change the literal value 5 to 10000.

The expression following the 'while' keyword is a condition that determines whether the statement block is executed or continues to execute. As long as the condition is true, the program executes the code block once more. When the condition becomes false, the loop is finished. If the condition is false initially, the code block within the body of the loop is not executed at all. The indented code block following the while keyword is often simply called the body.

The while statement has the general form:

```
while condition :
block
```

- The keyword while begins the while statement.
- The condition determines whether the body will be (or will continue to be) executed. A colon (:) must follow the condition.
- block is a block of one or more statements to be executed as long as the condition is true. As a block, all the statements that comprise the block must be indented the same number of spaces from the left. As with the 'if' statement, the block must be indented more spaces than the line that begins the while statement. The block is technically part of the while statement.

Except for the keyword while instead of 'if', while statements look identical to if statements. Sometimes beginning programmers confuse the two or accidentally type 'if' when they mean 'while' or vice versa. Usually the very different behavior of the two statements reveals the problem immediately; however, sometimes, especially in nested, complex logic, this mistake can be hard to detect.

Figure 5.1 shows how program execution flows through Listing 5.2 (iterativecounttofive.py).

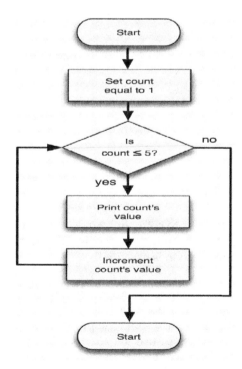

FIGURE 5.1
While flowchart.

The condition is checked before the body is executed, and then checked again each time after the block has been executed. If the condition is initially false the block is not executed. If the condition is initially true, the block is executed repeatedly until the condition becomes false, at which point the loop terminates.

Program 5.3 (addition.py) allows users to enter a positive integer value. When the user enters a negative integer the loop terminates and the code prints the sum of the positive integers entered up to that point. The sum is zero if the first figure is negative.

```
entry = 0
sum = 0
print ("Enter numbers to sum, negative number ends list:")
while entry >= 0:
  entry = eval (input ())
if entry >= 0:
  sum += entry
print ("Sum =", sum)
```

Output

```
Enter numbers to sum, negative number ends list:
45
-78
Sum = 0
```

Two variables, input, and sum in Program 5.3 (addition.py)

- **Variable 'entry'**
  In the beginning we initialize 'entry' to zero because we want the condition 'entry > = 0' of the 'while' statement to be true initially. If we fail to initialize 'entry', the program will produce a run-time error when it attempts to compare 'entry' to zero in the 'while' condition. 'entry' holds the number entered by the user and its value can change each timethe loop is executed.

- **Variable 'sum'**
  The variable 'sum' is known as an accumulator, because it accumulates each value the user enters. Clearly we must initialize 'sum' to zero. If we fail to initialize 'sum', the program generates a run-time error when it attempts to use the += operator to modify the variable. Within the loop we repeatedly add the user's input values to 'sum'. When the loop finishes (because the user entered a negative

number), 'sum' holds the sum of all the non-negative values entered by the user.

The initialization of 'entry' to zero coupled with the condition 'entry >= 0' guarantees that the body of the 'while' loop will execute at least once. The 'if' statement ensures that a negative entry will not be added to 'sum'. (Could the 'if' condition have used > instead of >= and achieved the same results?) When the user enters a negative value, 'sum' will not be updated and the condition of the 'while' will be false. The loop then terminates and the program executes the 'print' statement.

Program 5.3 (addition.py) shows that a 'while' loop can be used for more than simple counting. The program does not keep track of the number of values entered. It simply accumulates the entered values in the variable named 'sum'.

A 'while' statement can be used to make Program 4.10 (troubleshoot.py) more convenient for the user. Recall that the computer troubleshooting program forces the user to rerun the program once a potential fault has been detected (for example, turn on the power switch, then run the program again to see what else might be wrong). A better decision logic is shown in Figure 5.2.

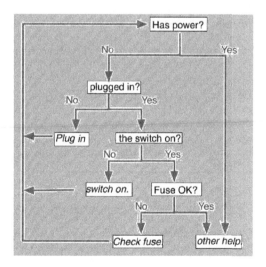

FIGURE 5.2
Troubleshooting decision tree.

Program 5.4 (troubleshootloop.py) incorporates a 'while' statement so that the program's execution continues until the problem is resolved or its resolution is beyond the capabilities of the program.

```
print("Help! My computer doesn't work")
print("Does the computer make any sounds (fans, etc.)")
choice = input("or show any lights? (y/n):")
if choice =='n':
   choice = input("Is it plugged in? (y/n):")
if choice =='n':
   print("plug it in. If the problem persists, ")
   print("please run this program again.")
else:
choice = input("Is the switch in the \"on\" position?
(y/n):")
if choice == 'n':
   print("Turn it on. If the problem persists,")
   print("please run this program again.")
else:
   choice = input("Does tje computer have a fuse? (y/n):")
if choice == 'n':
   choice = input("Is the outlet's OK? (y/n)")
   if choice == 'n':
      print("Check the outlet's circuit")
      print("btraker or fuse. Move to a")
      print("new outlet, if necessary.")
      print("If the problem persists")
      print("Please run this program again.")
else:
   print("Please consult a service technician.")
   '''else:
   print("Check the fuse. Replace if")
   print("necessary. If the problem")
   print("persists,then")
   print("Please run this program again.")
   else:
print("Please consult a service technician.")
```

Output

```
Help! My computer doesn't work
Does the computer make any sounds (fans, etc.)
or show any lights? (y/n):y
Is the switch in the "on" position? (y/n):y
Does the computer have a fuse? (y/n):y
Please consult a server technician.
```

Most rules or standards are wrapped in a declaration. The finished Boolean variable governs the loop and the loop continues until it is false. You can think of the flag being down when the value is false and raised when it is true. In this case, when the flag is raised, it is a signal that the loop should terminate [30].

5.2 Difference Between Definite and Indefinite Loops

In Program 5.5 (counttofive.py), the entries in the list are written from one to 10.

```
n = 1
while n <= 10:
    print(n)
    n += 1
```

Output

```
1
2
3
4
5
6
7
8
9
10
```

We can inspect the code and determine the number of iterations the loop performs. This kind of loop is known as a definite loop, since we can predict exactly how many times the loop repeats [31]. Consider Program 5.6 (def2.py).

```
n = 1
stop = int(input())
while n <= stop:
print(n)
```

Output

IDLE Shell 3.9.6   —   □   ✕

File Edit Shell Debug Options Window Help

```
Python 3.9.6 (tags/v3.9.6:db3ff76, Jun 28 2021, 15:26:21) [MSC v.1929 64 bit (AM
D64)] on win32
Type "help", "copyright", "credits" or "license()" for more information.
>>>
================ RESTART: C:/Users/Chanderbhan/Desktop/PP1.py ================
```

We cannot predict how many times the loop of Program 5.6 (definite2.py) will repeat. The number of iterations depends on the input provided by the user. However, at the program's point of execution after obtaining the user's input and before the start of the execution of the loop, we would be able to determine the number of iterations the 'while' loop would perform. The loop in Program 5.6 (definite2.py) is thus also considered to be a definite loop.

Compare these programs to Program 5.7 (indefinite.py).

```
done = False
while not done:
    entry = eval(input())
    if entry == 999:
        done = True
    else:
        print(entry)
```

Output

```
 "IDLE Shell 3.9.6"                                             —    □    ×

File   Edit   Shell   Debug   Options   Window   Help
Python 3.9.6 (tags/v3.9.6:db3ff76, Jun 28 2021, 15:26:21) [MSC v.1929 64 bit (AM
D64)] on win32
Type "help", "copyright", "credits" or "license()" for more information.
>>>
================ RESTART: C:\Users\Chanderbhan\Desktop\PP1.py ================
2|

1
1
1
1
1
1
1
1
1
1
1
1
1
1
```

In Program 5.7 (indefinite.py), we cannot predict at any point during the loop's execution how many iterations the loop will perform. The value to match (999) is known before and during the loop, but the variable entry can be anything the user enters. The user could choose to enter 0 exclusively or enter 999 immediately and be done with it. The 'while' statement in Program 5.7 (indefinite.py) is an example of an indefinite loop. Program 5.4 (troubleshootloop.py) is another example of an indefinite loop.

The 'while' statement is ideal for indefinite loops. Although we have used the 'while' statement to implement definite loops, Python provides a better alternative for definite loops: the 'for' statement.

5.3 The For Loop

In Python the 'while' loop is used where the programmer cannot anticipate how frequently a loop will be performed (i.e. an indefinite loop).

```
n = 1
while n <= 10:
print (n)
n += 1
```

Output

```
1
2
3
4
5
6
7
8
9
10
```

Python provides a different structure for definite loops, known as the 'for' loop. An argument with a definite set of values is iterated. The argument may be used directly or may refer to data structure elements such as sequences, lists, or tuples. The above loop can be rewritten using a 'for' loop thus:

```
for n in range(1, 11):
   print (n
```

The 'range' keyword, together with the bracketed values(1, 11) produces the entity 'n', which is the iterable. It takes the values 1, 2, ..., 10 and is used directly within the indented code block that follows the 'for' keyword. . The full definition of a 'for' loop is

```
for  iterable in range (begin, end, step):
```

where

- *begin* is the initial value of the iterable; 0 if omitted.
- *end* is the final value of the iterable; the end value cannot be omitted
- *step* is the increment applied to the iterable at each cycle of the loop, which may be positive or negative and is assumed to be plus one if omitted. All these parameters must be integers. Floating-point numbers are not permissible [32].

For example the following loop decrements in steps of 3 from 21 down to 3:

```
for n in range (21, 0, -3):
    print (n,   ", end=")
```

Output

21   18   15   12   9   6   3

The range(1000) gives the series 0,1,2, ... ,999.

The following python program sums all positive integers other than 100 and prints all the intermediate values:

```
sum = 0
for n in range (1, 100):
    sum = sum + n
    print ("sum", sum)
```

Output

sum 1
sum 3
sum 6
sum 10
sum 15
sum 21
sum 28
sum 36
sum 45
sum 55
sum 66
sum 78
sum 91

The following examples illustrate how several sequences may be generated:

- range (10) → 0,1,2,3,4,5,6,7,8,9
- range (1,10) → 1,2,3,4,5,6,7,8,9
- range (1,10,2) → 1,3,5,7,9
- range (10,0,-1) → 10,9,8,7,6,5,4,3,2,1
- range (10,0,-2) → 10,8,6,4,2
- range (2,11,2) → 2,4,6,8,10
- range (-5,5) → -5,-4,-3,-2,-1,0,1,2,3,4

- range (1, 2) → 1
- range (1, 1) → (empty)
- range (1, -1) → (empty)

5.4 Nested Loop Statements

As with 'if' statements, Python 'for' loops, can include other loops within them [33]. These are called 'nested loops'. Python program 5.8 (table.py) displays a multiplication table on the screen.

```
# Print a multiplication table to 10 x 10
# Print column heading
print("1     2     3     4     5     6     7     8     9
    10")
print("  +-------------------------------------------------")
for row in range(1, 11): # 1 <= row <= 10, table has 10
    rows
    if row < 10: # Need to add space?
        print(" ", end="")
    print(row, "| ", end="") # Print heading for this row.
    for column in range(1, 11): # Table has 10 columns.
        product = row*column; # Compute product
        if product < 100: # Need to add space?
            print(end=" ")
        if product < 10: # Need to add another space?
            print(end=" ")
        print(product, end=" ") # Display product
    print()
```

Output

	1	2	3	4	5	6	7	8	9	10
1	1	2	3	4	5	6	7	8	9	10
2	2	4	6	8	10	12	14	16	18	20
3	3	6	9	12	15	18	21	24	27	30
4	4	8	12	16	20	24	28	32	36	40
5	5	10	15	20	25	30	35	40	45	50
6	6	12	18	24	30	36	42	48	54	60
7	7	14	21	28	35	42	49	56	63	70
8	8	16	24	32	40	48	56	64	72	80
9	9	18	27	36	45	54	63	72	81	90
10	10	20	30	40	50	60	70	80	90	100

- It is important to distinguish what is done only once (outside all loops) from that which is done repeatedly. The column heading across the top of the table is outside of all the loops; therefore, it is printed once in the beginning.

- The work to print the heading for the rows is distributed throughout the execution of the outer loop. This is because the heading for a given row cannot be printed until all the results for the previous row have been printed.

```
A code fragment like
if x < 10:
      print(end=" ")
print(x, end=" ")
```

prints x in one of two ways: if x is a one-digit number, it prints a space before it; otherwise, it does not print the extra space. The net effect is to right justify one and two digit numbers within a two character space printing area. This technique allows the columns within the times table to be properly right aligned [34].

- row is the iterable for the outer loop, and column the iterable for the inner loop.

- The inner loop executes ten times on every single iteration of the outer loop. How many times is the statement

```
product = row*column    # Compute product
      executed? 10×10 = 100, once for every product in
      the table.
```

- A newline is printed after the contents of each row is displayed; thus, all the values printed in the inner (column) loop appear on the same line.

```
# Print a MAX x MAX multiplication table
MAX = 18
# First, print heading
print(end=" ")
# Print column heading numbers
for column in range(1, MAX + 1):
    print(end=" %2i " % column)
print() # Go down to the next line

 # Print line separator; a portion for each column
print(end=" +")
for column in range(1, MAX + 1):
    print(end="----") # Print portion of line
```

```
print() # Go down to the next line

# Print table contents
for row in range(1, MAX + 1): # 1 <= row <= MAX, table
has MAX rows
    print(end="%2i | " % row) # Print heading for this
    row.
    for column in range(1, MAX + 1): # Table has 10
    columns.
        product = row*column; # Compute product
        print(end="%3i " % product) # Display product
    print() # Move cursor to next row
```

Output

```
    1   2   3   4   5   6   7   8   9  10  11  12  13  14  15  16  17  18
   +--------------------------------------------------------------------
 1 |  1   2   3   4   5   6   7   8   9  10  11  12  13  14  15  16  17  18
 2 |  2   4   6   8  10  12  14  16  18  20  22  24  26  28  30  32  34  36
 3 |  3   6   9  12  15  18  21  24  27  30  33  36  39  42  45  48  51  54
 4 |  4   8  12  16  20  24  28  32  36  40  44  48  52  56  60  64  68  72
 5 |  5  10  15  20  25  30  35  40  45  50  55  60  65  70  75  80  85  90
 6 |  6  12  18  24  30  36  42  48  54  60  66  72  78  84  90  96 102 108
 7 |  7  14  21  28  35  42  49  56  63  70  77  84  91  98 105 112 119 126
 8 |  8  16  24  32  40  48  56  64  72  80  88  96 104 112 120 128 136 144
 9 |  9  18  27  36  45  54  63  72  81  90  99 108 117 126 135 144 153 162
10 | 10  20  30  40  50  60  70  80  90 100 110 120 130 140 150 160 170 180
11 | 11  22  33  44  55  66  77  88  99 110 121 132 143 154 165 176 187 198
12 | 12  24  36  48  60  72  84  96 108 120 132 144 156 168 180 192 204 216
13 | 13  26  39  52  65  78  91 104 117 130 143 156 169 182 195 208 221 234
14 | 14  28  42  56  70  84  98 112 126 140 154 168 182 196 210 224 238 252
15 | 15  30  45  60  75  90 105 120 135 150 165 180 195 210 225 240 255 270
16 | 16  32  48  64  80  96 112 128 144 160 176 192 208 224 240 256 272 288
17 | 17  34  51  68  85 102 119 136 153 170 187 204 221 238 255 272 289 306
18 | 18  36  54  72  90 108 126 144 162 180 198 216 234 252 270 288 306 324
```

In python program 5.9 (flexibletimestable.py) we use loops to vary how we print the headings. This program works just as well for 1×1 and 15×15 times tables. It is trivial to modify python program 5.9 (flexibletimestable. py) so that the size of the table printed is based on user input instead of a programmer-defined constant.

Nested loops are used when an iterative process itself must be repeated. Python program 5.9 (flexibletimestable.py) uses an inner 'for' loop to print the contents of each row, but multiple rows must be printed. The inner (column) loop prints the contents of each row, while the outer (row) loop is responsible for printing all the rows.

Python program 5.10 (permuteabc.py) uses a triple nested loop to print all the different arrangements of the letters A, B, and C. Each string printed is a permutation of ABC.

```
# File permuteabc.py

# The first letter varies from A to C
for first in 'ABC':
    for second in 'ABC': # The second varies from A to C
        if second != first: # No duplicate letters allowed
            for third in 'ABC': # The third varies from A to C
                # Don't duplicate first or second letter
                if third != first and third != second:
                    print(first + second + third)
```

Output

```
ABC
ACB
BAC
BCA
CAB
CBA
```

A quadruple nested loop is used in python program 5.11 (abcd.py) to print all the different possible arrangements of A, B, C and D. Each printed string is an ABCD permutation.

```
# File permuteabcd.py

# The first letter varies from A to D
for first in 'ABCD':
 for second in 'ABCD': # The second varies from A to D
  if second != first: # No duplicate letters allowed
   for third in 'ABCD': # The third varies from A to D
    # Don't duplicate first or second letter
    if third != first and third != second:
     for fourth in 'ABCD': #The fourth varies from A to D
      if fourth != first and fourth != second and fourth
      != third:
       print(first + second + third + fourth)
```

Output

```
ABCD
ABDC
ACBD
ACDB
ADBC
ADCB
BACD
BADC
BCAD
BCDA
BDAC
BDCA
CABD
CADB
CBAD
CBDA
CDAB
CDBA
DABC
DACB
DBAC
DBCA
DCAB
DCBA
```

5.5 Abnormal Loop Termination

Normally, a 'while' statement executes until its condition becomes false. Since this condition is checked only at the "top" of the loop, the loop is not immediately exited if the condition becomes false due to activity in the middle of the body. Ordinarily this behavior is not a problem because the intention is to execute all the statements within the body as an indivisible unit. Sometimes, however, it is desirable to immediately exit the body or recheck the condition during its execution. Python provides 'break' and 'continue' statements to give programmers more flexibility in designing the control logic of loops [35].

5.5.1 The 'Break' Statement

The break statement causes an immediate exit from the body of the loop. Python program 5.12 (addmiddleexit.py) is a variation of python program 5.3 (addnonnegatives.py) that illustrates the use of 'break'.

```
entry = 0 # Ensure the loop is entered
sum = 0 # Initialize sum

 # Request input from the user
print("Enter numbers to sum, negative number ends list:")

while True: # Loop forever
  entry = eval(input()) # Get the value
  if entry < 0: # Is number negative number?
    break # If so, exit the loop
  sum += entry # Add entry to running sum
print("Sum =", sum) # Display the sum
```

Output

```
Enter numbers to sum, negative number ends list:
-87
sum = 0
```

There are two examples in python program 5.12. In both the condition of the 'while' loop is always true, so the body of the loop will be entered. Since the condition can never be false, the break statement is the only way to get out of the loop. The break statement is executed only when the user enters a negative number. When the break statement is encountered during the program's execution, the loop is immediately exited. Any statements following the break within the body are skipped. In the second example it is not possible, therefore, to add a negative number to the sum variable.

A program may require a designated variable to monitor loop life. Python program 5.13 (troubleloop.py) uses the Boolean instead of split statements.

```
print("Help! My computer doesn't work!")
while True:
  print("Does the computer make any sounds (fans, etc.)")
  choice = input(" or show any lights? (y/n):")
  # The troubleshooting control logic
  if choice == 'n': # The computer does not have power
    choice = input("Is it plugged in? (y/n):")
    if choice == 'n': # It is not plugged in, plug it in
      print("Plug it in.")
    else: # It is plugged in
      choice = input("Is the switch in the \"on\"
      position? (y/n):")
      if choice == 'n': # The switch is off, turn it on!
```

```
            print("Turn it on.")
        else: # The switch is on
            choice = input("Does the computer have a fuse?
            (y/n):")
            if choice == 'n': # No fuse
                choice = input("Is the outlet OK? (y/n):")
                if choice == 'n': # Fix outletpr
                    int("Check the outlet's circuit ")
                    print("breaker or fuse. Move to a")
                    print("new outlet, if necessary. ")
                else: # Beats me!
                    print("Please consult a service technician.")
                    break # Nothing else I can do, exit loop
            else: # Check fuse
                print("Check the fuse. Replace if ")
                print("necessary.")
    else: # The computer has power
        print("Please consult a service technician.")
        break # Nothing else I can do, exit loop
```

Output

```
Help! My computer doesn't work!
Does the computer make any sounds (fans, etc.)
or show any light? (y/n): y
Please consult a service technician.
```

'Break' statements should be used sparingly as a deviation is inserted into the standard loop function. Ideally, each loop should have single input and output points. Split declarations are commonly used by the programmer for a specified series of conditions. The inclusion of a 'break' provides an additional escape point in such a 'while' loop (the start of theloop is one point, and the 'break' declaration is another). It is particularly undesirable to use multiple break statements within a single loop. In such a while loop, adding a break statement adds an extra exit point (the top of the loop where the condition is checked is one point, and the break statement is another). Using multiple break statements within a single loop is particularly dubious and should be avoided. Why have the break statement at all if its use is questionable and it is dispensable? The logic is fairly simple, so the restructuring is straightforward; in general, the effort may complicate the logic a bit and require the introduction of an additional Boolean variable. Any program that uses a break statement can be rewritten so that the break statement is not used. Any loop of the form.

5.5.2 'Continue' Statements

The 'continue' statement is similar to the 'break' statement. During a program's execution, when the break statement is encountered within the body of a loop, the remaining statements within the body of the loop are skipped, and the loop is exited. When a continue statement is encountered within a loop, the remaining statements within the body are skipped, but the loop condition is checked to see if the loop should continue or be exited. If the loop's condition is still true, the loop is not exited, but the loop's execution continues at the top of the loop. Python program 5.14 (continuestatement. py). illustrates how to use the continue statement.

```
sum = 0
done = False;
while not done:
   val = eval(input("Enter positive integer (999 quits):"))
   if val < 0:
     print("Negative value", val, "ignored")
     continue; # Skip rest of body for this iteration
   if val != 999:
     print("Tallying", val)
     sum += val
   else:
     done = (val == 999); # 999 entry exits loop
print("sum =", sum)
```

Output

```
Enter positive integer (999 quits): 45
Tallying 45
Enter positive integer (999 quits): 12
Tallying 12
Enter positive integer (999 quits): 100000
Tallying 100000
Enter positive integer (999 quits): 0
Tallying 0
Enter positive integer (999 quits):
```

The 'continue' statement is not used as often as the 'break' statement since equivalent code is more easily written.

The logic of the else version is no more complex than the continue version. Therefore, unlike the break statement above, there is no compelling reason to use the continue statement. Sometimes a continue statement is added at the last minute to an existing loop body to handle an exceptional condition (like ignoring negative numbers in the example above) that initially went unnoticed. If the body of the loop is lengthy, a conditional statement with a continue can be added easily near the top of the loop body without touching the logic of the rest of the loop. Therefore, the continue statement merely provides a convenient alternative to the programmer. The else version is preferred.

5.6 Infinite Looping Statement

An infinite loop executes its block of statements repeatedly until the user forces the program to quit. Once the program flow enters the loop's body it cannot escape. Infinite loops are sometimes designed. For example, a long-running server application like a Web server may need to continuously check for incoming connections. This checking can be performed within an infinite loop. However, infinite loops may be accidentally created by beginners..

Intentional infinite loops should be made obvious. For example,

```
while True:
    # Do something forever. . .
```

The Boolean literal 'True' is always true, so it is impossible for the loop's condition to be false. The only ways to exit the loop are through a 'break' statement, a 'return' statement (see Chapter 7), or a sys.exit call embedded somewhere within its body.

```
# List the factors of the integers 1..MAX
MAX = 20 # MAX is 20
n = 1 # Start with 1
while n <= MAX: # Do not go past MAX
    factor = 1 # 1 is a factor of any integer
    print(end=str(n) + ': ') # Which integer are we
    examining?
    while factor <= n: # Factors are <= the number
        if n % factor == 0: # Test to see if factor is a
        factor of n
            print(factor, end=' ') # If so, display it
        factor += 1 # Try the next number
    print() # Move to next line for next n
    n += 1
```

Accidental infinite loops, however, commonly confuse beginners. Find python program 5.14 (factors.py) attempting to print all entries from 1 to 20. It shows

```
1: 1
2: 1
3: 1
```

and then "freezes up" or "hangs," ignoring any user input (except the key sequence Ctrl-C on most systems which interrupts and terminates the running program). This type of behavior is a frequent symptom of an unintentional infinite loop. The factors of 1 display properly, as do the factors of 2. The first factor of 3 is properly displayed and then the program hangs. Since the program is short, the problem may be easy to locate. In some programs, though, the error may be challenging to find. Even in Listing 5.16 (findfactors.py) the debugging task is nontrivial since nested loops are involved. (Can you find and fix the problem in Listing 5.16 (findfactors.py) before reading further?)

In order to avoid infinite loops, we must ensure that the loop exhibits certain properties:

- The loop's condition must not be a tautology (a Boolean expression that can never be false). For example,

```
while i >= 1 or i <= 10:
    # Block of code follows ...
```

is an infinite loop since any value chosen for i will satisfy one or both of the two subconditions. Perhaps the programmer intended to use 'and' instead of 'or' to stay in the loop as long as i remains in the range 1…10.

In Listing 5.16 (findfactors.py) the outer loop condition is

```
n <= MAX
```

If n is 21 and MAX is 20, then the condition is false. Since we can find values for n and MAX that make this expression false, it cannot be a tautology. Checking the inner loop condition:

```
factor <= n
```

we see that if factor is 3 and n is 2, then the expression is false; therefore, this expression also is not a tautology.

- The condition of a 'while' must be true initially to gain access to its body. The code within the body must modify the state of the program in some way so as to influence the outcome of the condition that is checked at each iteration. This usually means one of the variables used in the condition is modified in the body. Eventually the variable assumes a value that makes the condition false, and the loop terminates.

  In Listing 5.16 (findfactors.py) the outer loop's condition involves the variables 'n' and 'MAX'. We observe that we assign 20 to 'MAX' before the loop and never change it afterward, so to avoid an infinite loop it is essential that n be modified within the loop. Fortunately, the last statement in the body of the outer loop increments 'n'. 'n' is initially 1 and 'MAX' is 20, so unless the circumstances arise to make the inner loop infinite, the outer loop eventually should terminate.

  The inner loop's condition involves the variables 'n' and 'factor'. No statement in the inner loop modifies 'n', so it is imperative that 'factor' be modified in the loop. Although 'factor' is incremented in the body of the inner loop, the increment operation is protected within the body of the 'if' statement. The inner loop contains one statement, the 'if' statement. That 'if' statement in turn has two statements in its body:

```
while factor <= n:
  if n % factor == 0:
    print(factor, end=' ') factor += 1
```

If the condition of the 'if' is ever false, the variable factor will not change. In this situation if the expression 'factor <= n' was true, it would remain true. This effectively creates an infinite loop. The statement that modifies factor must be moved outside the 'if' statement's body.

If the condition of the if is always false, the variable factor will not change. In this situation if the expression factor <= n was true, it will remain true. This effectively creates an infinite loop. The statement that modifies the factor must be moved outside of the if statement's body.

This current version executes correctly:

```
while factor <= n:
  if n % factor == 0:
    print(factor, end=' ')
factor += 1
```

```
1:  1
2:  1 2
3:  1 3
4:  1 2 4
5:  1 5
6:  1 2 3 6
7:  1 7
8:  1 2 4 8
9:  1 3 9
10: 1 2 5 10
11: 1 11
12: 1 2 3 4 6 12
13: 1 13
14: 1 2 7 14
15: 1 3 5 15
16: 1 2 4 8 16
17: 1 17
18: 1 2 3 6 9 18
19: 1 19
20: 1 2 4 5 10 20
```

A debugger can be used to step through a program and see where and why it is an endless loop. Another popular strategy is to put print statements in specific positions to evaluate the values of the loop control variables.

```
while factor <= n:
        print ('factor = ', factor, ' n =', n)
if n % factor == 0:
        print (factor, end=' ')
factor += 1    # <-- Note, still has original error here
```

The modified code above generates the following output:

```
1: factor = 1 n = 1
1
2: factor = 1 n = 2
1 factor = 2 n = 2
2
3: factor = 1
n = 3
1 factor = 2
```

```
n = 3
factor = 2
n = 3 factor = 2
   n = 3 factor = 2
n = 3
factor = 2 n = 3
factor = 2 n = 3
.
.
.
```

The program continues to print the same line until the user interrupts its execution. The output demonstrates that once 'factor' becomes equal to 2 and 'n' becomes equal to 3 the program's execution becomes trapped in the inner loop. Under these conditions:

1. 2 < 3 is true, so the loop continues and
2. 3 % 2 is equal to 1, so the 'if' statement will not increment 'factor'.

It is imperative that factor be incremented each time through the inner loop; therefore, the statement incrementing 'factor' must be unindented to move it outside the 'if's' guarded body. [37].

Python Program 5.17 (findfactorsfor.py) is a different version of our factor finder program that uses nested 'for' loops instead of nested 'while' loops. Not only is it slightly shorter, but it avoids the potential for the misplaced increment of the 'factor' variable. This is because the 'for' statement automatically handles the loop variable update

```
File  Edit  Format  Run  Options  Window  Help
# Print a MAX x MAX multiplication table
MAX = 18
# First, print heading
print(end=" ")
  # Print column heading numbers
for column in range(1, MAX + 1):
    print(end=" %2i " % column)
print() # Go down to the next line

  # Print line separator; a portion for each column
print(end=" +")
for column in range(1, MAX + 1):
    print(end="----") # Print portion of line
print() # Go down to the next line

  # Print table contents
for row in range(1, MAX + 1): # 1 <= row <= MAX, table has MAX rows
    print(end="%2i | " % row) # Print heading for this row.
    for column in range(1, MAX + 1): # Table has 10 columns.
        product = row*column; # Compute product
        print(end="%3i " % product) # Display product
    print() # Move cursor to next row
```

Output

```
     1   2   3   4   5   6   7   8   9  10  11  12  13  14  15  16  17  18
  +-------------------------------------------------------------------------------
  1 |  1   2   3   4   5   6   7   8   9  10  11  12  13  14  15  16  17  18
  2 |  2   4   6   8  10  12  14  16  18  20  22  24  26  28  30  32  34  36
  3 |  3   6   9  12  15  18  21  24  27  30  33  36  39  42  45  48  51  54
  4 |  4   8  12  16  20  24  28  32  36  40  44  48  52  56  60  64  68  72
  5 |  5  10  15  20  25  30  35  40  45  50  55  60  65  70  75  80  85  90
  6 |  6  12  18  24  30  36  42  48  54  60  66  72  78  84  90  96 102 108
  7 |  7  14  21  28  35  42  49  56  63  70  77  84  91  98 105 112 119 126
  8 |  8  16  24  32  40  48  56  64  72  80  88  96 104 112 120 128 136 144
  9 |  9  18  27  36  45  54  63  72  81  90  99 108 117 126 135 144 153 162
 10 | 10  20  30  40  50  60  70  80  90 100 110 120 130 140 150 160 170 180
 11 | 11  22  33  44  55  66  77  88  99 110 121 132 143 154 165 176 187 198
 12 | 12  24  36  48  60  72  84  96 108 120 132 144 156 168 180 192 204 216
 13 | 13  26  39  52  65  78  91 104 117 130 143 156 169 182 195 208 221 234
 14 | 14  28  42  56  70  84  98 112 126 140 154 168 182 196 210 224 238 252
 15 | 15  30  45  60  75  90 105 120 135 150 165 180 195 210 225 240 255 270
 16 | 16  32  48  64  80  96 112 128 144 160 176 192 208 224 240 256 272 288
 17 | 17  34  51  68  85 102 119 136 153 170 187 204 221 238 255 272 289 306
 18 | 18  36  54  72  90 108 126 144 162 180 198 216 234 252 270 288 306 324
```

5.7 Examples of Iteration

This section offers many examples of 'if' and 'while' statements to illustrate their use.

5.7.1 Computation of a Square Root

This Python program first calculates square root of the user's number. It then squares the result to see how close to the original number the result is. It quits if the result is very close to the right one.

```python
# File computesquareroot.py

# Get value from the user
val = eval(input('Enter number: '))
 # Compute a provisional square root
root = 1.0;

# How far off is our provisional root?
diff = root*root - val
```

```
# Loop until the provisional root
# is close enough to the actual root
while diff > 0.00000001 or diff < -0.00000001:
    root = (root + val/root) / 2 # Compute new provisional
    root
    print(root, 'squared is', root*root) # Report how we
    are doing
    # How bad is our current approximation?
    diff = root*root - val

# Report approximate square root
print('Square root of', val, "=", root)
```

A basic algorithm is used to provide a response that is within 0.00000001 of the correct answer with consecutive zero approximations.

Output

```
Enter number: 2
1.5   squared is 2.25
1.4166666666666665 squared is 2.006944444444444
1.4142156862745097 squared is 2.0000060073048824
1.4142135623746899 squared is 2.0000000000045106
Square root of 2 = 1.4142135623746899
```

The real root of the square is roughly 1.4142135623730951 and thus our threshold (0.00000001) has been acknowledged.

```
Enter number: 100
50.5   squared is 2550.25
26.24009900990099 squared is 688.542796049407
15.025530119986813 squared is 225.76655538663093
10.840434673026925 squared is 117.51502390016438
10.032578510960604 squared is 100.6526315785885
10.000052895642693 squared is 100.0010579156518
10.000000000139897 squared is 100.00000000279795
Square root of 100 = 10.000000000139897
```

Naturally, the true solution is ten, but our measured result is still within our configured tolerance.

Whilst list 5.18 (squareroot.py) is an outstanding illustration of how to utilize a loop, Python has a library feature that is more reliable and effective if we need to calculate a square root. It and other practical mathematical functions are investigated.

5.7.2 Structure of Tree Drawing

Assume that we want to sketch a geometric tree, and the user supplies the height. A five-level tree that appears thus

```
        *
       ***
      *****
     *******
    *********
```

A three-level tree looks like this.

```
      *
     ***
    *****
```

The program can be written as a simple amendment to python program chapter 1 (arrow.py), which uses no loops. However, our program, based on user input, needs to differ in height and width.

The necessary functionality is python program 5.17 (starconcept.py). When another python program 5.17 (starconcept.py) is executed & also a number of user enter, the results are:

```
# Get tree height from user
height = eval(input("Enter height of tree: "))

# Draw one row for every unit of height
row = 0
while row < height:
  # Print leading spaces; as row gets bigger, the number
  of
  # leading spaces gets smaller
  count = 0
  while count < height - row:
    print(end=" ")
    count += 1
    # Print out stars, twice the current row plus one:
    # 1. number of stars on left side of tree
    # = current row value
    # 2. exactly one star in the center of tree
    # 3. number of stars on right side of tree
    # = current row value
```

```
count = 0
while count < 2*row + 1:
    print(end="*")
    count += 1
# Move cursor down to next line
print()
row += 1 # Consider next row
```

Output

```
Enter height of tree: 5
        *
       ***
      *****
     *******
    *********
```

- As long as the user enters a value greater than zero, the body of the outer 'while' loop will be executed; if the user enters zero or less, the program terminates and does nothing.
- The last statement in the body of the outer 'while':

  ```
  row += 1
  ```

  ensures that the variable row increases by one each time through the loop; therefore, it eventually will equal 'height' (since it initially had to be less than 'height' to enter the loop), and the loop will terminate. There is no possibility of an infinite loop here.

The two inner loops play distinct roles:

- The first inner loop prints spaces. The number of spaces printed is equal to the height of the tree the first time through the outer loop and decreases for each iteration. This is the correct behavior since each succeeding row moving down contains fewer leading spaces but more asterisks.
- The second inner loop prints the row of asterisks that make up the tree. The first time through the outer loop, row is zero, so no left side asterisks are printed, one central asterisk is printed (the top of the tree), and no right side asterisks are printed. Each time through the loop the number of left-hand and right-hand stars to print both increase by one and the same central asterisk is printed; therefore, the tree grows one wider on each side each line moving down.

While it seems asymmetrical, note that no third inner loop is required to print trailing spaces on the line after the asterisks are printed. The spaces would be invisible, so there is no reason to print them. Python program 5.18 (star1concept.py) uses our star-trees 'for' loops instead of loops.

```python
# Get tree height from user
height = eval(input("Enter height of tree: "))
# Draw one row for every unit of height
for row in range(height):
# Print leading spaces; as row gets bigger, the number of
# leading spaces gets smaller
for count in range(height - row):
print(end=" ")
for count in range(2*row + 1): print(end="*")
# Move cursor down to next line
print()
```

Output:

```
Enter height of tree: 7
              *
              *
              *
              *
              *
              *
              *
            * * *
            * * *
            * * *
            * * *
            * * *
            * * *
          * * * * *
          * * * * *
          * * * * *
          * * * * *
          * * * * *
        * * * * * * *
        * * * * * * *
        * * * * * * *
        * * * * * * *
      * * * * * * * * *
      * * * * * * * * *
      * * * * * * * * *
    * * * * * * * * * * *
    * * * * * * * * * * *
  * * * * * * * * * * * * *
```

5.8 Program to Print Prime Numbers

Prime numbers are integers that have no integer factors except themselves and 1.For instance, 29 is a prime number, but 28 (2, 4, 7, and 14 are factors of 28) is not a prime number. Prime numbers were once a mathematician's intellectual interest, but are now of importance in cryptography.

The goal of this python program is to display all the prime numbers up to a user inserted value.

```
max_value = eval(input('Display primes up to what value?
'))
value = 2 # Smallest prime number
while value <= max_value:
  # See if value is prime
  is_prime = True # Provisionally, value is prime
  # Try all possible factors from 2 to value - 1
  trial_factor = 2
  while trial_factor < value:
      if value % trial_factor == 0:
        is_prime = False; # Found a factor
        break # No need to continue; it is NOT prime
      trial_factor += 1 # Try the next potential factor
  if is_prime:
    print(value, end= ' ') # Display the prime number
  value += 1 # Try the next potential prime number
print() # Move cursor down to next line
```

Output

```
Display primes upto what value? 89
2   3   5   7   11   13   17   19   23   29   31   37   41
43   47   53   59   61   67   71   73   79   83   89
```

This progam is a bit more complicated than python program 5.19 (printconcept.py). The max value is given by the user. The outer loop iterates for all values from two to max:

- The attribute is key, which means it's the highest value unless otherwise indicated by our studies. Both values of 2 to -1 are taken in the internal phase trial factor:

```
trial_factor = 2
while trial_factor < value:
if value % trial_factor == 0:
        is_prime = False;        # Found a factor
break      # No need to continue; it is NOT prime
   trial_factor += 1        # Try the next potential factor
```

If a trial factor is separated into interest and no residual, the sum percent trial factor is zero – specifically if the trial factor is a number. If all of the trial factor values are found to be an important factor, then its prime is set to zero, and the process is escaped by the interruption. When the process ends, its prime can never be set to zero, which implies that no variables have been established or meaning is key.

- Whether declaration during the internal loop:

```
if is_prime:
    print(value, end= ' ') # Display the prime number
```

Only test the is prime level. If is prime is valid then the value will be primary, and the value may be written, with the next iteration being accompanied by a separating space from other variables.

There might be any critical issues.

- Will it be written if the consumer enters a 2,?
  Max value = value = 2 for the state of the external loop

  ```
  value <= max_value
  ```

  Because 2 vs 2 is real. That is valid. is prime is set to correct, but the internal loop state

  ```
  trial_factor < value
  ```

  Not real (2 is just 2). Not valid. The internal loop has then been removed, is prime is not modified, and 2 is written. This is the right behavior, as 2 is the least (and the only equal) first) number.

- Was something written if the client has fewer than 2 entries,?
  It implies that fewer than two meanings are not taken into account. It would never move into the body of the victim. It is written only

on the newline and numbers are not included. It's there the best behavior.

- Is the internal loop always meant to end?
  The trial factor must be lower than the value to join the internal body of the ring. No change of interest in the process anywhere. Trial factor in the if the statement is not changed anywhere on the list, and is raised directly after the if declaration inside the list. Therefore, a trial factor is raised any time the process is iterated. Finally, the trial factor receives the same meaning and the process stops.

- Is the external loop still expected to end?
  The value must be smaller than or equal to the max value to join the core of the outer loop. Nothing in the loop shifts max value. In the final declaration in the external loop structure, the value is raised and no other value is altered. When the internal loop continues as seen ultimately, the value hits max value in the previous statement, which ends the loop.

  To stop breaking the argument, the rationale of the inner thought maybe reorganized somewhat. This version is as follows:

```
while trial_factor < value:
if value % trial_factor == 0: is_prime = False; # Found a
factor
break   #      No need to continue; it is NOT prime
trial_factor += 1    #     Try the next potential factor
It can be rewritten as:
while is_prime and trial_factor < value:
is_prime = (value % trial_factor != 0) # Update is_prime
trial_factor += 1
```

This non-break variant is a little more complex by extracting the statement while in the body. The primary is set to reality after the loop.. It is reassigned any time around the process. If at any point the meaning percent trial factor becomes 0, the trial factor would become wrong. That is exactly when the trial factor is a beneficial element. If the loop is wrong, the loop may not start, even if the loop never stops, if the loop is equal to the meaning of the trial factor. Thanks to the order of the server, the brackets are

```
is_prime = (value % trial_factor != 0)
```

They're not wanted. The parentheses boost the readability as both = and! = people are difficult to examine. If parenthesis is put where it is not essential, as in

```
x = (y + 2);
```

The translator lacks them, and the running program does not provide an output cost.

The Description python program (printprimesnumber.py) as above may be simplified a little by rendering assertions instead of definitions as stated python program 5.20 (print.py).

```
max_value = eval(input('Display primes up to what value?
')) # Try values from 2 (smallest prime number) to
max_value
for value in range(2, max_value + 1): # See if value is
prime
is_prime = True # Provisionally, value is prime # Try all
possible factors from 2 to value - 1
for trial_factor in range(2, value):
if value % trial_factor == 0:
is_prime = False    # Found a factor
break   # No need to continue; it is NOT prime
if is_prime:
print(value, end= ' ') # Display the prime number
print() # Move cursor down to next line
```

5.8.1 Inputs

Python program 5.21 (meaning.py) tests the number input by a user to ensure it is within a valid range.

```
# Require the user to enter an integer in the range 1-10
in_value = 0 # Ensure loop entry
attempts = 0 # Count the number of tries

# Loop until the user supplies a valid number
while in_value < 1 or in_value > 10:
  in_value = int(input("Please enter an integer in the
  range 0-10: "))
  attempts += 1

# Make singular or plural word as necessary
tries = "try" if attempts == 1 else "tries"
# in_value at this point is guaranteed to be within range
print("It took you", attempts, tries, "to enter a valid
number")
```

Output

```
Please enter an integer in the range 0-10: 11
Please enter an integer in the range 0-10: 1
It took you 2 tries to enter a valid number
```

We initialise the 'value' variable to 0 at the beginning of the program, to ensure that the body runs at least once.

5.9 Conclusion

The looping constructs 'while' and 'for' allow sections of code to be executed repeatedly under some condition. The statement "'for' iterates over a fixed range of values or a sequence. The statements within the body of the 'for' loop are executed till the range of values is exhausted. The statements within the body of a 'while' loop are executed repeatedly until the condition of the 'while' becomes false. If the condition of the while loop is initially false, the body is never executed. The statements within the body of the 'while' loop must ensure that the condition eventually becomes false; otherwise, the loop will become infinite, because of a logical error in the program. The 'break'" statement immediately exits a loop, skipping the rest of the loop's body. Execution continues with the statement immediately following the body of the loop. When a 'continue' statement is encountered, the control jumps to the beginning of the loop for the next iteration.

6

Functions

6.1 Introduction to Functions

Large programs are often difficult to manage and so are divided into smaller units known as functions [38]. A function is simply a group of statements assigned to a certain name, i.e. the function name, and can be invoked (called) from another part of the program. For example, School Management software will contain various tasks like Registering Student, Fee Collection, Library Book Issue, TC Generation or Result Declaration. In this case we have to create different functions for each task to maintain the steps of software development. A set of functions is stored in a file called a "module." This approach is known as "modularization" and makes a program easier to understand, test and maintain. Commonly used modules that contain source code for generic needs are called "libraries." Functions are of mainly two types.

6.1.1 Built-in Functions

Built-in functions are the ready-made functions in Python that are frequently used in programs. Let us inspect the following Python program.

```
#Program to calculate square of a number
a = int(input("Enter a number: ")
b = a * a
print(" The square of ",a ,"is", b)
```

In the above program, input(), int() and print() are the built-in functions. The set of instructions to be executed for these built-in functions are already defined in the Python interpreter.

Now let us consider the following Python statement consisting of a function call to a built-in function and answer the given questions.

```
fname = input("Enter your name: ")
```

What is the name of the function being used?
input()
Does the function accept both a value and an argument?
Yes, because the parenthesis "()" consists of a string "Enter your name".
Does the function return a value?
Yes, since there is an assignment (=) operator preceding the function name, the function returns a value which is stored in the variable fname.
Hence, the function input () accepts a value and returns a value.
Now consider the built-in functions int () and print ().
The following is a categorized list of some of the frequently used built-in functions in Python.

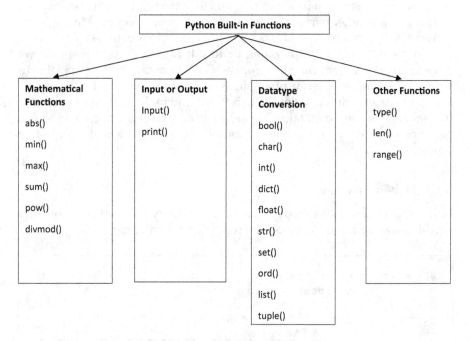

6.1.2 User-defined Functions

A function definition begins with def (short for "define"). The syntax for creating a user defined function is:

```
def<Function name> ([parameter 1, parameter 2,....]): Function Header
        set of instructions to be executed
        [return <value>]
```

Function Body (Should be indented within the function header)

The items enclosed in "[ ]" are called parameters and they are optional. Hence, a function may or may not have parameters. A function may or may not return a value.

A function header always ends with a colon (:).

The function name should be unique. The rules for naming identifiers also applies for function naming.

The statements outside the function indentation are not considered to be part of the function.

- A function with no arguments and no return: This type of function is also known as a "void function."

```
def pwelcome():
    print("hello Python Environment")
    print("start python")
```

Output

```
pwelcome()
hello Python Environment
start python
```

- A function with arguments but no return value: Parameters are given in the parenthesis separated by comma. Values are passed from the parameter at the time of function calling.

```
def table(num):
    for i in range(1, 11):
        print(num, "x",i,"=",num*i)
n = int(input("Enter Any Number"))
table(n)
```

Output

```
Enter Any Number 10
10 x 1 = 10
10 x21 = 20
10 x 3 = 30
10 x 4 = 40
10 x 5 = 50
10 x 6 = 60
10 x 7 = 70
10 x 8 = 70
10 x 9 = 90
10 x 10 = 100
```

- A function with arguments and a return value:

```
def average(num1, num2, num3, num4):
  av = (num1+num2+num3+num4) / 4.0
  print("Average Value =",av)

n1=int(input("Enter First Number "))
n2=int(input("Enter Second Number "))
n3=int(input("Enter Third Number "))
n4=int(input("Enter Fourth Number "))
average(n1,n2,n3,n4)
```

Output

```
Enter First Number 4
Enter Second Number 4
Enter Third Number 4
Enter Fourth Number 4
Average Value = 4.0
```

- A function with no arguments but a return value:

```
import math
def area(r):
  a = math.pi * r * r
  return a
n = int(input("Enter Radius "))
ar = area(n)
print("Area of Circle = ",ar)
```

Output

```
Enter Radius 4
Area of Circle =  50.26548245743669
```

6.2 The Meaning of a Function

Function definition starts with a def keyword followed by the function name and the arguments the function can take. These arguments are within parentheses and there can be any number of arguments including zero. When there is more than one argument, they are separated by a comma. Since Python is a loosely typed language (meaning we don't assign a data type to a variable when declaring), we shouldn't include the data types for the arguments. The function definition statement is terminated by a colon. In Python, we define a function as follows [39]:

- **Defining the function**: The function description includes the code for the actions of the function and defines the feature.
- **Invocation function**: In a program, a method is activated to connect the invoking functions. We will use normal functions in this chapter, which will not be specified by us. There is one description for increasing functions; however, several invocations may occur.

There are three sections of an ordinary function definition:

- **Name**: There's a description in most Python functions. The name is an ID (see Section 2.3). The name selected for a feature will, like variable names, correctly reflect its intended intent or functionality. (Python enables the lambda function to be programmed openly. This function is the same as a regular Python function but can be defined without a name.)
- **Parameters**: The parameters it accepts from callers are defined in each role specification. The parameters appear in a comma-separated table. If the function does not need any code details that call the element, the parameter list is zero.
- **Body**: The block of indented statements which constitute a function body is present in each function definition. When the client contacts the device, the body contains the code to be run. The code concept is very helpful and we can change the structure of code according to client need.

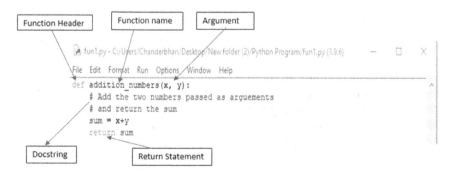

6.3 Documenting Functions

Functions are like mini-programs within a program and group a bunch of statements so that they can be used and reused throughout different parts of the program [40].

Python Function-related Statements with Code Example

Statements	Sample Code Example
def, parameters, return	def add(a, b=1, *args, **kwargs): return a + b + sum(args) + sum(kwargs.values())
calls	add(3,4,5, 9, c=1, d=8) # Output: 30

Most of us find it hard to document our functions as it is time-consuming and boring. However, while not documenting our code, in general, may seem all right for small programs, when the code gets more complex and large, it will be hard to understand and maintain.

This section is to encourage us to always document our functions, no matter how small our programs may seem to be.

6.3.1 Importance of Documenting a Function

There is a saying that "Programs must be written for people to read, and only incidentally for machines to execute."

We can't stress enough that documenting our functions helps other developers (including ourselves) to easily understand and develop our code.

We may come across code we wrote years ago and say: "What was I thinking?" This is because there was no documentation to remind us of what the code did, and how it did it.

That being said, documenting our functions or code, in general, brings the following advantages:

- Adds more meaning to our code, thereby making it clear and understandable.
- Eases maintainability. With proper documentation, we can return to our code years later and still be able to modify it swiftly.
- Eases development. In an open-source project, for example, many developers work on the codebase simultaneously. Poor or no documentation will discourage such development of our projects.
- Enables popular integrated development environment (IDE) debugging tools to effectively assist us in our development.

6.3.2 Documenting Functions with Python Docstrings

According to PEP 257 Docstring Conventions: "A docstring is a string literal that occurs as the first statement in a module, function, class, or method definition. Such a docstring becomes the __doc__ special attribute of the object."

Docstrings are defined with the **triple-double quote** (""") string format. At a minimum, a Python docstring should give a quick summary of whatever the function is doing.

A function's docstring can be accessed in two ways. Either directly via the function's **__doc__** special attribute or using the built-in help() function which accesses **__doc__** behind the hood.

Example 6.1　Access a function's docstring via the function's __doc__ special attribute.

```
def add(a, b):
    """Return the sum of two numbers(a, b)"""
    return a + b

if __name__ == '__main__':
    print(add.__doc__)
```

Output

```
Return the sum of two numbers(a, b)
```

The docstring above represents a **one-line** docstring. It appears in one line and summarizes what the function does.

Example 6.2　Access a function's docstring using the built-in help() function.

Run the following command from a Python shell terminal.

```
Python 3.9.6 (tags/v3.9.6:db3ff76, Jun 28 2021, 15:26:21)
  [MSC v.1929 64 bit (AMD64)] on Win32
Type "help", "copyright", "credits" or "license()" for
  more information.
>>> help(sum) # access docstring of sum()
Help on built-in function sum in module builtins:
```

```
sum(iterable, /, start=0)
    Return the sum of a 'start' value (default: 0) plus
    an iterable of numbers
    When the iterable is empty, return the start value.
    This function is intended specifically for use with
    numeric values and may
    reject non-numeric types.
```

Press **q** to exit this display.

A multi-line Python docstring is more thorough, and may contain all the following:

- The function's purpose;
- Information about arguments;
- Information about return data;
- Any other information that may seem helpful to us.

Example 6.3 shows a thorough way of documenting our functions. It starts by giving a short summary of what the function does, and a blank line followed by a more detailed explanation of the function's purpose, then another blank line followed by information about arguments, the return value and exceptions, if any.

Example 6.3

```
def add_ages(age1, age2=30):
    """ Return the sum of age
    Sum and return the ages of your son and daughter

    Parameters
    -----------
    age1: int
        The age of your son
    age2: int, Optional
        The age of your daughter(default to 30)
    Return
    ----------
    age : int
        The sum of your son and daughter ages.
    """
    age = age1 + age2
    return age
```

```
if __name__ == '__main__':
    print(add_ages.__doc__)
```

Notice the break-space after the enclosing triple-quote before the function's body.

Output

```
Return the sum of ages
    Sum and return the ages of your son and daughter

    Parameters
    -----------
        age1: int
            The age of your son
        age2: int, Optional
            The age of your daughter(default to 30)
    Return
    -----------
        age : int
            The sum of your son and daughter ages.
```

Note: This is not the only way to document using docstring. Read on for other formats too.

6.3.3 Python Docstring Formats

The docstring format used above is the NumPy/SciPy-style format. Other formats also exist. We can also create our format to be used by our company or open-source. However, it is good to use well-known formats recognized by all developers.

Some other well-known formats are Google docstrings, reStructuredText and Epytext.

6.4 GCD Function

The greatest common divisor or gcd is a mathematical expression to find the highest number which can divide two numbers with a remainder of zero. It has many mathematical applications. Python has an inbuilt gcd function in the math module which can be used for this purpose [41]. It accepts two integers as parameters and returns the integer which is the gcd value.

Syntax
```
Syntax: gcd(x,y)
Where x and y are positive integers.
```

Example of gcd()

Here we print the result of the gcd of a pair of integers.

```
import math
print("GCD of 75 and 30 is ", math.gcd(75, 30))
print("GCD of 0 and 12 is ", math.gcd(0, 12))
print("GCD of 0 and 0 is ", math.gcd(0, 0))
print("GCD of -24 and -18 is ", math.gcd(75, 30))
```

Output

```
GCD of 75 and 30 is 15
GCD of 0 and 12 is 12
GCD of 0 and 0 is 0
GCD of -24 and -18 is 6
```

In the above example, the `math.gcd()` function generates the gcd of two given numbers. In the `gcd()` function, a and b pass an argument that returns the greatest common divisor of two integer numbers, completely dividing the numbers.

6.5 The Main Function

The main function is like the entry point of a program. However, the Python interpreter runs the code right from the first line. The execution starts from the starting line and proceeds line by line. It does not matter whether the main function is present or not.

Since there is no `main()` function in Python, when the command to run a program is given to the interpreter, the code that is at level 0 indentation is executed. However, before doing that, it will define a few special variables. __name__ is one such special variable. If the source file is executed as the main program, the interpreter sets the __name__ variable to have a value __main__. If this file is being imported from another module, __name__ will be set to the module's name.

__name__ is a built-in variable which evaluates to the name of the current module [42].

```
# Python program to demonstrate
# main() function

print("Hello")
# Defining main function
def main():
          print("hey there")

# Using the special variable
#  __name__
if __name__ =="__main__":
        main ()
```

Output

```
Hello
hey there
```

When the above program is executed, the interpreter declares the initial value of name as "main". When the interpreter reaches the if statement it checks the value of name and if it is true it runs the main function; otherwise the main function is not executed.

6.6 The Calling Function

As we know, functions are a block of statements used to perform some specific tasks in programming. It also helps to break a large group of code into smaller chunks or modules. Functions can be called anywhere and any number of times in a program. This allows us to reuse the code by simply calling the particular function or block in a program. Thus, it avoids the repetition of the same code. We can define functions inside a class, module, nested function, and so on.

Features of functions
- Used to avoid repetitions of code;
- Using the function, we can divide a group of code into smaller modules;
- Helps to hide the code and create clarity to understand modules;
- Allows code to be reused, thus saving memory;
- Statements written inside a function can only be executed with a function name;
- A Python function starts with def and then a colon (:) followed by the function name.

Rules for defining a function

- The def keyword is used in the function to declare and define it;
- The function name must begin with the following identifiers: A–Z, a–z or underscore (_);
- Every function must follow a colon (:) and then use indention to write the program;
- Reserved words cannot be used as a function name or identifier;
- In Python, the function parameter can be empty or have multiple entries.

Creating a function in Python

To create a function, we need to use a def keyword. Here is the syntax:

- def function_name(): # use def keyword to define the function;
- Statement to be executed;
- return statement # return a single value.

Let's create a function program.

```
def myFun(): # define function name
print(" Welcome to Python Envirinment")
myFun() # call to print the statement
```

Output

```
Welcome to Python Environment
```

Function calling in Python

Once a function is created in Python, we can call it by writing function_name () itself or another function/nested function. The following is the syntax for calling a function.

```
   def function_name():
   Statement1
function_name() # directly call the function
   # calling function using built-in function
def function_name():
str = function_name('john') # assign the function to call
the function
print(str) # print the statement
```

Consider the following example to print a Welcome Message using a function in Python.

```
def MyFun():
    print("Hello World")
    print(" Welcome to the JavaTpoint")
  MyFun() # Call Function to print the message.
```

Output

```
Hello World
Welcome to the JavaTpoint
```

In the above example, we call the MyFun() function that prints the statements.

Calling nested functions in Python

When we construct one function inside another, it is called a nested function. We can create nested functions using the def keyword. After creating the function, we have to call the outer and the inner function to execute the statement. Let's create a program to understand the concept of nested functions and how we can call them.

Nest.py

```
def OutFun():
  print("Hello, it is the outer function")

  def InFun():
    print("Hello, It is the inner function")
  InFun()

OutFun()
```

Output

```
Hello, it is the outer function
Hello, it is the inner function
```

As we can see in the above example, the InFun() function is defined inside the OutFun() function. To call the InFun() function, we first call the OutFun() function in the program. After that, the OutFun() function will start executing and then call InFun() as the above output.

6.7 Argument Passing in Parameters (Actual and Formal)

6.7.1 Parameters vs. Arguments

Parameters are the names used when defining a function or a method, and into which arguments will be mapped. In other words, arguments are the things which are supplied to any function or method call, while the function or method code refers to the arguments by their parameter names.

Consider the following example and look back at the above DataCamp Light chunk: you pass two *arguments* to the sum() method of the Summation class, even though you previously defined three *parameters*, namely, self, a and b.

What happened to self?

The first argument of every class method is always a reference to the current instance of the class, which in this case is Summation. By convention, this argument is called self.

This all means that you don't pass the reference to self in this case because self is the parameter name for an implicitly passed argument that refers to the instance through which a method is being invoked. This is inserted implicitly into the argument list.

How to define a function: user-defined functions (UDFs)

The four steps to defining a function in Python are:

- Use the keyword def to declare the function and follow this up with the function name.
- Add parameters to the function: they should be within the parentheses of the function; end your line with a colon.
- Add statements that the function should execute.
- End your function with a return statement if the function should output something; without the return statement, your function will return an object "None."

6.7.2 Function Arguments in Python

Earlier, we learned about the difference between parameters and arguments. In short, arguments are the things which are given to any function or method call, while the function or method code refers to the arguments by their parameter names. There are four types of arguments that Python UDFs can take:

Default arguments: These take a default value if no argument value is passed during the function call. You can assign this default value by the assignment operator (=), just as in the following example:

```
def plus(a,b = 2):
  return a + b

plus(a=1)

plus(a=1, b=3)
```

Output

```
IDLE Shell 3.9.6                                          —    □    ×
File  Edit  Shell  Debug  Options  Window  Help
Python 3.9.6 (tags/v3.9.6:db3ff76, Jun 28 2021, 15:26:21) [MSC v.1929 64 bit (AM
D64)] on win32
Type "help", "copyright", "credits" or "license()" for more information.
>>>
================= RESTART: C:\Users\Chanderbhan\Desktop\PP1.py =================
>>> 3 + 5
8
```

Required arguments: As the name suggests, the required arguments of a UDF are those that have to be there. These arguments need to be passed during the function call and in precisely the right order, just as in the following example:

```
# Define 'plus()' with required arguments
def plus(a,b):
    return a + b
```

You need arguments that map to the a as well as the b parameters to call the function without getting any errors. If you switch around a and b, the result won't be different, but it might be if you change plus () to the following:

```
script.py
1    # Define 'plus()' with required arguments
2▾   def plus(a,b):
3        return a/b
```

Keyword arguments: If you want to make sure that you call all the parameters in the right order, you can use the keyword arguments in your function
call. You use these to identify the arguments by their parameter name. Let's
take the example from above to make this a bit more clear:

```
script.py
1    # Define 'plus()' function
2▼   def plus(a,b):
3        return a + b
4
5    # Call 'plus()' function with parameters
6    plus(2,3)
7
8    # Call 'plus()' function with keyword arguments
9    plus(a=1, b=2)
```

Note that by using the keyword arguments, you can also switch around the
order of the parameters and still get the same result when you execute your
function:

```
script.py
1    # Define "plus()' function
2▼   def plus(a,b):
3        return a + b
4
5    # Call "plus()' function with keyword arguments
6    plus(b=2, a=1)
```

Variable number of arguments: In cases where you don't know the exact
number of arguments that you want to pass to a function, you can use the
following syntax with `*args`:

```
script.py
1    # Define 'plus()' function to accept a variable
     number of arguments
2    def plus(*args):
3        return sum(args)
4
5    # Calculate the sum
6    plus(1,4,5)
```

The asterisk (*) is placed before the variable name that holds the values of all non-keyword variable arguments. Note here that you might as well have passed *varint, *var_int_args or any other name to the plus () function.

We can try replacing *args with another name that includes the asterisk. You'll see that the above code keeps working!

You can see that the above function makes use of the built-in Python sum() function to sum all the arguments that get passed to plus(). If you would like to avoid this and build the function entirely yourself, you can use this alternative [43].

6.7.3 Global vs. Local Variables

In general, variables that are defined inside a function body have a local scope, and those defined outside have a global scope. That means that local variables are defined within a function block and can only be accessed inside that function, while global variables can be obtained by all functions that might be in your script:

```
script.py
1    # Global variable 'init'
2    init = 1
3
4    # Define 'plus()' function to accept a variable
     number of arguments
5▾   def plus(*args):
6        # Local variable 'sum()'
7        total = 0
8▾       for i in args:
9            total += i
10   return total
11
12   # Access the global variable
13   print("this is the initialized value " + str(init))
14
15   # (Try to) access the local variable
16   print("this is the sum " + str(total))
```

You'll see that you get a NameError that says that the name 'total' is not defined when you try to print out the local variable total that was defined inside the function body. The init variable, on the other hand, can be printed out without any problems.

6.7.4 Anonymous Functions in Python

Anonymous functions are also called lambda functions in Python because instead of declaring them with the standard `def` keyword, you use the `lambda` keyword.

```
script.py
1    double = lambda x: x*2
2
3    double(5)
```

In the DataCamp Light chunk above, `lambda x: x*2` is the anonymous or lambda function. x is the argument, and x*2 is the expression or instruction that gets evaluated and returned. What's special about this function is that it has no name, like the examples that we have seen in the first part of this chapter Section 6.1.1). If you had to write the above function in a UDF, the result would be the following:

```
def double(x):
  return x*2
```

Let's consider another example of a lambda function where you work with two arguments:

```
script.py
1    # 'sum()' lambda function
2    sum = lairbda x, y: x + y;
3
4    # Call the 'sum()' anonymous function
5    sum(4,5)
6
7    # "Translate" to a UDF
8▼   def sum(x, y):
9         return x+y
```

You use anonymous functions when you require a nameless function for a short period of time, and which is created at runtime. Specific contexts in which this would be relevant is when you're working with `filter()`, `map()` and `reduce()`:

```
script.py
1    from functools import reduce
2
3    my_list = [1,2,3,4,5,6,7,0,9,10]
4
5    # Use lambda function with 'filter()'
6    filtered_list = list(filter(lambda x: (x*2 > 10), my_
     list ))
7
8    # Use lambda function with 'map()'
9    mapped_list = list(map(lambda x: x*2, my_list))
10
11   # Use lambda function with 'reduce()'
12   reduced_list = reduce(lambda x, y: x+y, my_list)
13
14   print(filtered_list)
15   print(mapped_list)
16   print(reduced_list)
```

The filter() function filters, as the name suggests, the original input list my_list on the basis of a criterion >10. With map(), on the other hand, you apply a function to all items of the list my_list. In this case, you multiply all elements by 2.

Note that the reduce() function is part of the functools library. You use this function cumulatively to the items of the my_list list, from left to right, and reduce the sequence to a single value, 55, in this case.

6.8 The Return Statement and Void Function

Return statements allow us to come out of a function and also return a value back to the caller of the function. They cause a function to stop its execution and hand over the value to whichever other statement called it. Even if you were to write some code after a return statement, it wouldn't get executed because control breaks out of the function.

Return statements help in getting information back from a function. But why would you need information back from a function? Because, while coding an actual application, you need to exchange the returned values back and forth between different parts of your programs. Functions like this are called **void**, and they return "None," Python's special word for "nothing". Here's an example of a void function:

The following examples demonstrate the use of the return statement in a function.

```
def first_example (sentence):
        return sentence
print(first_example('This is my first return statement'))
```

6.9 Scope of Variables and Their Lifetimes

The scope is nothing but the life of a variable within the program, and lifetime is nothing but the duration over which variables exist.

Local variables inside a function in Python:

- In the local variable, we will declare the variable inside the function;
- Here funt is the function name;
- x is the variable.

For example:

```
def funt()://it is function
x='hello local variable'//variable inside function
print(x)
funt()//here it is function call
```

Output

```
hello local variable
```

Variables in function parameters in Python:

- In the parameterizing variable, we pass the variable inside the function.
- We pass the message inside the function declaration. That message we will get as output.

For example:

```
def funt(x)://here 'x' is parameter for this function
print(x)
funt("hello parameter variable")//message passed inside
function declearation
```

Output

```
hello parameter variable
```

Global variables in Python:

- A global variable is defined in the main body;
- It can access any part of the program;
- The variable is declared outside the function.

For example:

```
x="welcome to global variable"
def funt():
print(x)//it is calling inside the function
funt()

print(x) //it is calling outside the function
```

Output

```
welcome to global variable
welcome to global variable
```

Scope of a nested function in Python:

- A function defined inside another function is called a nested function;
- It follows last in last out (LIFO) structure.

For example:

```
x= "first global"
def funt ( ):
x="secocnd global"
print(x)
```

```
funt ()
print(x)
```

Output

```
second global
first global
```

In the above output, we can observe LIFO in the function.

Two messages passed inside the function in two same-name variables have this output.

6.10 Function Examples

6.10.1 Function to Generate Prime Numbers

Write a program to generate a list of all the prime numbers that are less than 20.

Before starting it is important to note what a prime number is:

- A prime number has to be a positive integer;
- Divisible by exactly 2 integers (1 and itself);
- 1 is not a prime number.

While there are many different ways to solve this problem, here are a few different approaches.

Trick 1: For Loops

```
# Initialize a list
primes = []for possiblePrime in range(2, 21):

    # Assume number is prime until shown it is not.
    isPrime = True
    for num in range(2, possiblePrime):
        if possiblePrime % num == 0:
            isPrime = False

    if isPrime:
        primes.append(possiblePrime)
```

If you look at the inner `for loop`, notice that as soon as `isPrime` is False, it is inefficient to keep on iterating and would be more efficient to exit the loop.

```
# Initialize a list
primes = []for possiblePrime in range(2, 21):

    # Assume number is prime until shown it is not.
    isPrime = True
    for num in range(2, possiblePrime):
        if possiblePrime % num == 0:
            isPrime = False

    if isPrime:
        primes.append(possiblePrime)
```

Output

```
primes
[2, 3, 5, 7, 11, 13, 17, 19]
```

Trick 2: For Loops Using Break
Trick 2 is more efficient than Trick 1 because as soon as you find a given number isn't a prime number you can exit the loop using `break`.

```
# Initialize a list
primes = []
for possiblePrime in range(2, 21):

    # Assume number is prime until shown it is not.
    isPrime = True
    for num in range(2, possiblePrime):
        if possiblePrime % num == 0:
            isPrime = False
            break

    if isPrime:
        primes.append(possiblePrime)
```

Trick 3: For Loop, Break and Square Root
Trick 2 benefited from not doing unnecessary iterations in the inner `for loop`. Trick3 is similar except for the inner `range` function. Notice that this function is now `range(2, int(possiblePrime ** 0.5) + 1)`.

```
# Initialize a list
primes = []
for possiblePrime in range(2, 21):
```

```
# Assume number is prime until shown it is not.
isPrime = True
for num in range(2, int(possiblePrime ** 0.5) + 1):
    if possiblePrime % num == 0:
        isPrime = False
        break

if isPrime:
    primes.append(possiblePrime)
```

To explain why this approach works, it is important to note a few things. A composite number is a positive number greater than 1 that is not prime (which has factors other than 1 and itself). Every composite number has a factor less than or equal to its square root. For example, in the image of the factors of 15, notice that the factors in red are just the reverse of the green factors. In other words, by the commutative property of multiplication, 3 x 5 = 5 x 3. You just need to include the green pairs to be sure that you have all the factors [44].

6.10.2 Command Interpreter

Python is an interpreter language. This means it executes the code line by line. Python provides a Python shell, which is used to execute a single Python command and display the result.

It is also known as REPL (Read, Evaluate, Print, Loop), where it reads the command, evaluates the command, prints the result, and loops back to read the command again.

To run the Python shell, open the command prompt or power shell on Windows and terminal window on mac, write python and press **enter**. A Python prompt comprising three greater-than symbols (>>>) appears, as shown below

6.10.3 Restricted Input

Program (input.py) forces the user to enter a value within a specified range. We now easily can adapt that concept to a function. It uses a function named get_int_in_a range that does not return until the user supplies a proper value.

```
def get_int_in_range(first, last):
  # If the larger number is provided first,
  # switch the parameters
  if first > last:
    first, last = last, first
    # Insist on values in the range first...last
    in_value = int(input("Please enter values in the
    range " \
        + str(first) + "..." + str(last) + ": "))
    while in_value < first or in_value > last:
      print(in_value, "is not in the range", first, "...",
      last)
      in_value = int(input("Please try again: "))
    # in_value at this point is guaranteed to be within
    range
    return in_value;

  # main
  # Tests the get_int_in_range function
def main():
  print(get_int_in_range(10, 20))
  print(get_int_in_range(20, 10))
  print(get_int_in_range(5, 5))
  print(get_int_in_range(-100, 100))

main() # Run the program
```

Output

```
Please enter values in the range 10...20: 4
4 is not in the range 10 ... 20
Please try again: 21
21 is not in the range 10 ... 20
Please try again: 10
10
None
None
```

- In many programs, this functionality could be useful (see Program 6.1 (input .py)).
- Criteria are defined for high and low values. This helps the feature be more robust as it can be seen in a whole new context somewhere in the system and is always running properly.
- The function needs to be called as the first parameter and with the second parameter as the higher number. The feature would also ignore the parameters and adjust them as planned automatically:

```
num = get_int_in_range(20, 50)
```

works exactly like

```
num = get_int_in_range(50, 20)
```

- The Unpleasant input variable is being used to avoid twice the Boolean statement checked (to test it will be printed on the incorrect input line and to determine if the cycle will continue).

6.10.4 Die Rolling Simulation

Program 6.2 (rollingdie.py) is reorganized to function.

```
from random import randrange

# Roll the die three times
for i in range(0, 3):
    # Generate random number in the range 1...6
    value = randrange(1, 6)
    # Show the die
    print("+-------+")
    if value == 1:
        print("|       |")
        print("|   *   |")
        print("|       |")
    elif value == 2:
        print("| *     |")
        print("|       |")
        print("|     * |")
```

```
      elif value == 3:
         print("| * |")
         print("| * |")
         print("| * |")
      elif value == 4:
         print("| * * |")
         print("| |")
         print("| * * |")
      elif value == 5:
         print("| * * |")
         print("| * |")
         print("| * * |")
      elif value == 6:
         print("| * * * |")
         print("| |")
         print("| * * * |")
      else:
         print(" *** Error: illegal die value ***")
      print("+-------+")
```

Output

```
+-------+
| ** |
| * |
| ** |
+-------+
+-------+
| * |
| |
| * |
+-------+
+-------+
| ** |
| * |
| ** |
+-------+
```

The key function is vague on the specifics of the pseudo-allocation number creation in Program 6.2 (rollingdie.py). Moreover, the leader does not draw the die. Such core elements of the system, which can also be developed independently of the primary, have purposes.

Notice how the outcome of the roll call is moved to show the die explicitly as an argument:

```
show_die (roll ())
```

6.10.5 Tree Drawing Function

Program 6.12 (treedraw.py) is a tree drawing function in Python.

```
def tree(height):
 row = 0 # First row, from the top, to draw
 while row < height: # Draw one row for every unit of
 height
  # Print leading spaces
  count = 0
  while count < height - row:
   print(end=" ")
   count += 1
  # Print out stars, twice the current row plus one:
  # 1. number of stars on left side of tree
  # = current row value
  # 2. exactly one star in the center of tree
  # 3. number of stars on right side of tree
  # = current row value
  count = 0
  while count < 2*row + 1:
   print(end="*")
   count += 1
  # Move cursor down to next line
  print()
  # Change to the next row
  row += 1

# main
# Allows users to draw trees of various heights
def main():
 height = int(input("Enter height of tree: "))
 tree(height)

main()
```

Output

```
Enter height of tree: 1
```

Note that the nominal height is in the principal and the formal parameter in the tree as the local variable. There is no disagreement here because the two heights are two different numbers. The statement

```
tree(height)
```

is a parameter in the main height and the name happens to coincide with the formal parameter. The call function binds the main value to a tree-named tree formal parameter. The interpreter will monitor the height of each of the templates.

6.10.6 Floating-Point Equality

The number of a floating point does not comprise actual figures; it is final and interpreted internally as a discrete mantissa and exponent. As 1/3 cannot finally be represented in the decimal number system (base 10), 1/10 cannot precisely be represented by a fixed-numbered binary (base 2) number system. Frequently, there are no issues with this imprecision because software algorithms are developed using floating points and have to carry out precise calculations, for example by guiding a spaceship to a faraway world. Even minor errors can lead to total failures in such cases. Estimates of floating points can and must be implemented securely and effectively, and not without due consideration.

```
def main():
  x = 0.9
  x += 0.1
  if x == 1.0:
   print("OK")
  else:
   print("NOT OK")

main()
```

Output

```
OK
```

Consider Program (floataddition.py) for increasing our trust in floating-point numbers, which introduces two such numbers with double exact significance and tests for a specified meaning.

All is well calculated by the action. Program (badfloat.py) helps to monitor the loop with a floating-point number, with double precision:

```
def main():
 # Count to ten by tenths
 i = 0.0
 while i != 1.0:
  print("i =", i)
  i += 0.1

main()
```

Output

```
i = 26.200000000000102
i = 26.300000000000104
i = 26.400000000000105
i = 26.500000000000107
i = 26.600000000000108
i = 26.70000000000011
i = 26.80000000000011
i = 26.900000000000112
i = 27.000000000000114
i = 27.100000000000115
i = 27.200000000000117
i = 27.300000000000118
```

We anticipate it to cease when a loop variable exists, but unless Ctrl-C type is used the program is interrupted. In this example we are adding 0.1, but a problem now exists. Because 0.1 cannot exactly be represented under double precision fixed-point-representation constrictions, the repetitive adding of 0.1 leads to time-consuming errors. While 0.1 + 0.9 can equate to 1, 10-fold added 0.1 can be 1.000001 or 0.99999, not precisely 1 for sure.

This sequence of = = and is seen by the file example (badfloatcheck.py)! = The operators of the floating-point attribute are of dubious interest as contrasted. The best method is to test that two floating-point values are equally similar, which implies that they vary very little. By considering two floating points x and y, the absolute value of the discrepancy between < 0.00001 will be calculated. We can build an equal function and integrate the features. This function is provided by Program (floatequals1.py).

```
from math import fabs

# The == operator is checked first since some special
# floating-point values such as floating-point infinity
# require an exact equality check.
```

```
def equals(a, b, tolerance):
  return a == b or fabs(a - b) < tolerance;
# Try out the equals function
def main():
  i = 0.0
  while not equals(i, 1.0, 0.0001):
    print("i =", i)
    i += 0.1

main()
```

Output

```
i = 0.0
i = 0.1
i = 0.2
i = 0.3000000000000004
i = 0.4
i = 0.5
i = 0.6
i = 0.7
i = 0.7999999999999999
i = 0.8999999999999999
```

The third parameter, called tolerance, relates to how similar the first two parameters have to be to the same value. For certain unique floating-point values, such as infinity floating-point representation, the = = operator is used, so that the method governs = = even equality. Because Python uses Boolean operators for logical OR short-circuit evaluation, the more extensive test is not done if the = = operator means consistency.

When comparing two floating points to equality values, you should use a function as equal.

6.11 Arguments Passed by Reference Value

Every statement is passed to a parameter in a function when a function with arguments is invoked. Since all data in Python are items, an object variable is indeed an object description. The reference point for each statement is transferred to the parameter if you use a function with arguments. This is called pass-by-value in the definitions of programming. To make it simple, we say that when invoking a function, the value of an argument is passed to a parameter. The value is a description of the object. If the reasoning is an amount or a chain, no matter what alterations in the component inside of the function, the argument will be changed.

Program (Incrementpay.py)

```python
def main():
    x = 1
    print("Before the call, x is", x)
    increment(x)
    print("After the call, x is", x)

def increment(n):
    n += 1
    print("\tn inside the function is", n)

main()
```

Output

```
Before the call, x is 1
        n inside the function is 2
After the call, x is 1
```

The valuation of x(1) is passed, as shown in the output, to provoke the increase function (line 4). In line 8 (function) the parameter n is increased by 1 but x is not altered regardless of the function.

This is because statistics and string instruments are known as unchanging objects. It is impossible to change the material of immutable images. When another new number is assigned to a parameter, the new number is assigned to the variable and a new object is assigned to the attribute.

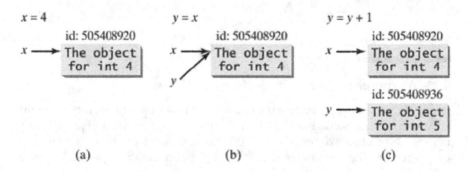

(a)   (b)   (c)

FIGURE 6.1

(a) 4 is assigned to x; (b) x is assigned to y; (c) y + 1 is assigned to y.

Consider the code below:

```
>>> x = 4
>>> y = x
>>> id(x)  # The reference of x
505408920
>>> id(y)  # The reference of y is the same as the
reference of x
505408920
```

Now x is assigned to y, while x and y now relate to the very same numeric 4 objects, as illustrated in Figure 6.1(a–b). But when adding 1 to y, a new object is created and assigned to. Now, as shown in the sample structure, y = + 1 # points to a new int object with value 5; and a new int object with value 5

```
>>> id (y)
505408936
```

6.12 Recursion

Recursion is the process of defining something in terms of itself. A physical world example would be to place two parallel mirrors facing each other. Any object in between them would be reflected recursively.

In Python, we know that a function can call other functions. It is even possible for the function to call itself. These types of construct are termed "recursive functions."

The following image shows the working of a recursive function called recurse.

```
def recurse():
    ...
    recurse() ──── recursive call
    ...

recurse() ────
```

The following is an example of a recursive function to find the factorial of an integer. The factorial of a number is the product of all the integers from 1 to that number. For example, the factorial of 6 (denoted as 6!) is 1*2*3*4*5*6 = 720.

Example of a recursive function

```
# Factorial of a number using recursion

def recur_factorial(n):
 if n == 1:
  return n
 else:
  return n*recur_factorial(n-1)

num = 7

# check if the number is negative
if num < 0:
 print("Sorry, factorial does not exist for negative
numbers")
elif num == 0:
 print("The factorial of 0 is 1")
else:
 print("The factorial of", num, "is", recur_factorial(num))
```

Output

```
The factorial of 7 is 5040
```

In the above example, factorial() is a recursive function since it calls itself.

When we call this function with a positive integer, it will recursively call itself by decreasing the number.

Each function multiplies the number with the factorial of the number below it until it is equal to one. This recursive call can be explained in the following steps.

```
factorial(3)          # 1st call with 3
3 * factorial(2)      # 2nd call with 2
3 * 2 * factorial(1)  # 3rd call with 1
3 * 2 * 1             # return from 3rd call as number=1
3 * 2                 # return from 2nd call
6                     # return from 1st call
```

Let's look at an image that shows a step-by-step process of what is going on:

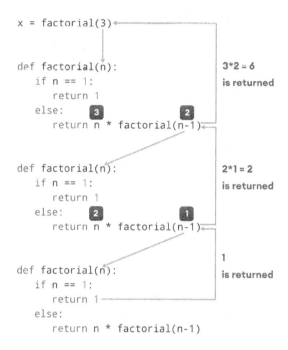

Working of a recursive factorial function

Our recursion ends when the number reduces to 1. This is called the base condition. Every recursive function must have a base condition that stops the recursion or else the function calls itself infinitely.

The Python interpreter limits the depths of recursion to help avoid infinite recursions, resulting in stack overflows.

By default, the maximum depth of recursion is 1000. If the limit is crossed, it results in RecursionError. Let's look at one such condition.

```
def recursor():
    recursor ()
recursor()
```

Output

```
== RESTART: C:/Users/Chanderbhan/Desktop/New folder (2)/Python
Program/fun1.py =
Traceback (most recent call last):
```

```
 File "C:/Users/Chanderbhan/Desktop/New folder (2)/Python
 Program/fun1.py", line 3, in <module> recursor()
 File "C:/Users/Chanderbhan/Desktop/New folder (2)/Python
 Program/fun1.py", line 2, in recursor recursor()
 File "C:/Users/Chanderbhan/Desktop/New folder (2)/Python
 Program/fun1.py", line 2, in recursor recursor()
 File "C:/Users/Chanderbhan/Desktop/New folder (2)/Python
 Program/fun1.py", line 2, in recursor recursor()
 [Previous line repeated 1022 more times]
RecursionError: maximum recursion depth exceeded
```

Advantages of recursion:

- Recursive functions make the code look clean and elegant;
- A complex task can be broken down into simpler sub-problems using recursion;
- Sequence generation is easier with recursion than using a nested iteration.

Disadvantages of recursion:

- Sometimes the logic behind recursion is hard to follow through;
- Recursive calls are expensive (inefficient) as they take up a lot of memory and time;
- Recursive functions are hard to debug.

6.13 Default Arguments

Python has a different way of representing syntax and default values for function arguments. Default values indicate that the function argument will take that value if no argument value is passed during the function call [45]. The default value is assigned by using the assignment (=) operator of the form keywordname=value. Let's understand this through a function stu-dent. This function contains three arguments of which two are assigned with default values. So, the function accepts one required argument (first-name), and the other two arguments are optional.

```
def student(firstname, lastname ='Mark', standard ='Fifth'):
   print(firstname, lastname, 'studies in', standard,
   'Standard')
```

We need to keep the following points in mind while calling functions:

1. In the case of passing the keyword arguments, the order of arguments is important;
2. There should be only one value for one parameter;
3. The passed keyword name should match with the actual keyword name;
4. In the case of calling a function containing non-keyword arguments, the order is important.

6.14 Time Functions

There are different types of functions linked to time in the time set.

The clock function helps one to calculate the time of execution of components of a system. The clock returns a floating-point value that reflects the period spent in seconds. Programs like Unix (Mac OS X and Linux) are used to calculate the number of seconds from the start of the application. Within Microsoft Windows, the clock provides the first application to control the number of seconds. In both instances, we can calculate elapsed time with two calls to the clock function. Program (timeit.py) tests the amount of time a person requires to enter a keyboard character. The following demonstrates the relationship between the program and a particularly sluggish typist.

```
from time import clock
print("Enter your name: ", end="")
start_time = clock()
name = input()
lapsed = clock() - start_time
print(name, "it took you", elapsed, "seconds to respond")
```

Output

```
Enter your name: Rick
Rick it took you 7.246477029927183 seconds to respond
```

The time taken for a program from Python to connect all the integers is computed in Program (timeaddition.py).

```
from time import clock
sum = 0 # Initialize sum accumulator
   start = clock() # Start the stopwatch
   for n in range (1, 100000001}: #Sum the numbers
   sum += n
elapsed = clock() - start #Stop the stopwatch
   print ("sum:", sum, "time:", elapsed) # Report results
```

Output

```
sum: 5000000050000000
time: 20.663551324385658
```

Program (measuretimepeed.py) tests the duration of the system to count up
to 100,000 primes using the same formula.

```
from time import clock
max_value = 10000 count = 0 start_time = clock ()      #
Start timer
# Try values from 2 (smallest prime number) to max_value
for value in range (2, max_value + 1): # See if value is
prime
   is prime = True # Provisionally, value is prime
   # Try all possible factors from 2 to value - 1
for trial_factor in range (2, value):
if value % trial_factor == 0:
is_prime = False        # Found a factor
   break # No need to continue; it is NOT prime
if is_prime:
            count += 1      # Count the prime number
print() # Move cursor down to next line
elapsed = clock () - start_time # Stop time timer
print ("Count:", count, " Elapsed time:", elapsed, "sec")
```

In one method, on average, all prime numbers up to 10,000 took about 1.25
seconds. In contrast, the algorithm in Program (moreefficientprimes.py)
which is based on the square root optimization is about ten times quicker
than in Program (moreefficientprimes.py). Accordingly the machine learns
the exact utility of the program.

```
from math import sqrt from time import clock
max_value = 10000 count = 0 value = 2    # Smallest prime
number
start = clock() # Start the stopwatch
while value <= max_value: # See if value is prime
   is_prime = True # Provisionally, value is prime
   # Try all possible factors from 2 to value - 1
trial_factor = 2 root = 3qrt(value)
while trial_factor <= root:
if value % trial_factor == 0:
is_prime = False;      # Found a factor
   break # No need to continue; it is NOT prime trial_
factor += 1
   # Try the text potential factor
```

Within an even quicker primary generator can be found; an entirely different method is used to produce prime numbers.

For a certain amount of seconds, the sleep feature suspends software execution.

The import time sleep for count in range(10, -1, -1): # Range 10, 9, 8, ..., 0 print(count) # Display the count sleep(1) # Suspend execution for 1 second. For quick testing of powerful visual motion graphics, the sleep feature is helpful.

6.15 Random Functions

Certain networks are forced to be unpredictable. In games or simulations, random variables are especially useful. Other board games use a die as an indicator to determine a player's number of places. For such casual games, a die or set of dice is used. A player occasionally plays a die roll or a pair of dice and their side(s) in the player matter. The full price after a transfer is calculated arbitrarily by the dynamic fall of the dying person. The simulation of the arbitrary dice roll of Figure 6.2 will be required for rolls-induced application modification [46].

These random number generators are algorithmic and will not necessarily generate pseudorandom numbers. There is a fixed time for a pseudorandom number generator depending on the design of the algorithm used. The sequence of the number generated reproduces itself exactly when worked long enough. A set of random numbers contains no replicated random

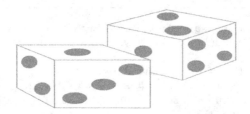

FIGURE 6.2
Pair of dice.

TABLE 6.1

A Few Functions of the Random Package

random
    Returns the floating-point pseudorandom integer x in 0 to x < 1
randrange
    Returns a pseudo-aligned integer in a given set
seed
    Sets the number of random seeds

---

numbers of this kind. The good news is that all working algorithmic generators have reasonably broad schedules for most applications.

The random Python module includes several basic functions and can be
used by programmers to deal with pseudorandom numbers. Table 6.1 displays each of these features.

The seed function determines the initial value of the pseudorandom
sequence numbers. Throughout the series of pseudo-allocated values, each
unique or unique request returns the next value. Program (simplerandom.
py) produces 100 pseudo-abandon integers in the 1 to 100 range.

```
from random import randrange, seed
seed(23)                                  #   Set random number
                                              seed
for i in range(0, 100):                   #   Print 100 random
                                              numbers
    print(randrange(1, 1000), end=")#     Range 1... 1,000
print()                                   #   Print newine
```

The program's numbers tend to be random. The seed value is given to begin
the algorithm, and the next value is generated by a formulation. The seed

value specifies the series of generated numbers; the seed values are similar. The same series is shown as the same number of seeds, 23, which will be used as the program is run again. The value of the seed must be specific for each run to allow each system to display various sequences.

The initial sequence value is dependent on the system's timing when we skip the call to seed. It is typically appropriate for basic pseudo-allocation numbers. This is helpful to define a seed value during design and testing if we want to see reproducible outcomes for software executions. Now we have everything we need to compose a program simulating the roll of a coin. Program (die.py) simulates rolling a die.

```
from random import randrange
# Roll the die three times
for i in range (0 , 3):
# Generate random number in the range 1 ... 6
value = randrange (1, 6)
# Show the die
print ("+-------+") if value == 1:
print ("|        |")
print ("|    *      |")
print ("|        |")
elif value == 2:
print ("|  *     |")
print ("|        |")
print ("|      * |")
elif value == 3:
print ("|      * |")
print ("|  *          |")
print ("| *     |")
elif value == 4:
          print ("| *      *  |")
print ("|        |")
print ("| *    *  |")
elif value == 5:
          print ("| *      *  |")
          print ("|      *          |")
          print ("| *      *  |")
elif value == 6:
          print ("| *  *  *  |")
          print ("|        |")
          print ("| *  *  * |") else:
print (" *** Error: illegal die value ***")
print ("+-------+")
```

Output

The actual performance differs from start to finish because the values are generated pseudorandomly.

6.16 Reusable Functions

In a function call, it is possible to use package functions in several different locations in the program. However, we have not seen so far how to define functions of certain applications that can be quickly reused. For instance, a prime function is functioning well inside and it could be included with some other primacy-testing programs (for example, the encryption framework uses high prime numbers). In order to copy the summary, the copying and pasting feature of our favorite text editor should be used [45].

It reuses a functionality even if the function's specification does not contain programmer-defined global variables or other programmer-defined parameters. When a function requires some of those external entities specified by the compiler, they still need to be used for a viable role in the new code. In other terms, the code in the specification of the feature

Only local parameters and variables are used. It is a truly unique function that can be conveniently repeated in other programs.

However, in principle it is not desirable that source code is transferred from one system to another. It is too easy to make the copy incomplete, or to make some other error while copying. Repetition of this programming is a waste of time. If 100 service provider programs are all required to use the primary function, all must use the primary software in the strategy. This redundancy wastes space. Last but not least, if a flaw is discovered in the prime function that contains all 100 programs, then in the most convincing example of the vulnerability of this copy-and-paste approach, the remaining 99 systems also include bugs as an error is detected and corrected in one system. You have to update your source code and it can be hard to determine which files you have to fix. If the application was made open to the general public, the issue is even greater. All of the copies of the defective function may not be tracked and corrected. If the correct function is prime and was updated to be more effective, the situation would be the same. The problem is that all programs that use primes define their function as prime; while function definitions are meant to be the same, all of those common definitions are not linked together. We want to recreate the functionality as it is without cloning it. Python allows developers to quickly bundle the functionality into modules. The modules can be provided separately from the applications since the functions are used for normal math and arbitrary modules. See Program (primenumrcode.py).

```
# Contains the definition of the is_prime function
from math import sqrt
# Returns True if non-negative integer n is prime;
# otherwise, returns false
def is_prime(n): trial_factor = 2 root = sqrt(n)
while trial_factor <= root:
if n % trialFactor == 0:# Is trialFactor a factor?
   return False; # Yes, retirn right away
         return True;# Tried them all, must he prime
from primecode import is_prime
def main():
num = int(input("Enter an integer: ")) if is_prime(num):
print(num, "is prime")
else:
print(num, "is NOT prime")
```

Code can be found in many Python programs in the Description package. This module exists in the simplest case in the same directory (folder) as that of the application computer package. Example Program (primenumcode.py) provides client sample software used in our package as a prime function.

The prime method is now simpler for many programs to use. When all the users of the program want to commonly use our Program (primecode.py) package the application will be put in a separate Python library folder and made accessible on the network to all users.

6.17 Mathematical Functions

Most of the mathematical calculator functions are supported by the basic math module. A few features are mentioned in Table 6.2.

In the math bundle, the values pi (π) and e (e) are described. The client's transferred parameter is referred to as the real parameter. The function parameter is formally named parameter. The first parameter of the actual format is assigned to the first formal parameter during the call function. When calling a feature, callers ought to be vigilant to place the claims they transfer in the correct sequence. The pow(10,2), but pow(2,10) counts $2^{10} = 1.024$. Name pow(10,2).

The math module needs to be imported into a Python system with some other mathematical features.

The problem can be addressed with the functions in the mathematical module in Figure 6.3. A satellite is some distance from earth at a set spot. A spacecraft is in a circular orbit around the earth. We want to determine how far the satellite is from the earth when its orbital direction has advanced 10 degrees.

Our coordinate structure (0,0), which is also in the middle of the circular orbital direction, is at the center of the planet. The satellite is at the beginning (x1,y1), then at stage (px, py); the spacecraft is stationary. The spacecraft is positioned on the same point as the orbit of the rocket. We have to measure two separate periods during the satellite's flight: the distance between both

---

TABLE 6.2

A Few Functions of the Math Package

mathfunctions Module

sqrt	Computes the square root of a number: sqrt $(x) = \sqrt{x}$
exp	Computes e raised a power: exp $(x) = e^x$
log	Computes the natural logarithm of a number: log $(x) = \log_e x = \ln x$
log10	Computes the common logarithm of a number: log $(x) = \log_{10} x$
cos	Computes the cosine of a value specified in radians: cos $(x) = \cos x$; other trigonometric functions include sine, tangent, arc cosine, arc sine, arc tangent, hyperbolic cosine, hyperbolic sine, and hyperbolic tangent
pow	Raises one number to a power of another: pow $(x,y) = x^y$
degrees	Converts a value in radians to degrees: degrees $(x) = \dfrac{\pi}{180} x$
radians	Converts a value in degrees to radians: radians $(x) = \dfrac{180}{\pi} x$
fabs	Computes the absolute value of a number: fabs $(x) = |x|$

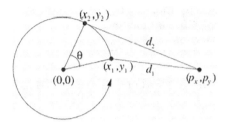

FIGURE 6.3
Orbital distance problem.

the ground station and the fixed point (space probe) and between both traveling points.

Two things need to be settled and mathematical facts answered:

1. Problem: While going around the loop, the traveling point will be recalibrated.

   Solution: With the first location (x_1,y_1) of the point of change, the rotation of an oscillating direction around a source creates a new point (x_2,y_2).

$$x_2 = x_1 \cos\theta - y_1 \sin\theta$$
$$y_2 = x_1 \sin\theta - y_1 \cos\theta$$

2. Problem: When the moving point enters a new location, it will recalculate the difference between the point of movement and the fixed point.

   Solution: Distance d1 is shown by formulation in Double dot px, py and x1,y1 Figure 6.2

$$d_1 = \sqrt{\left(x_1 - p_x\right)^2 + \left(y_1 - p_y\right)^2}$$

Likewise, in Figure 6.3, distance d2 is

$$d_2 = \sqrt{\left(x_2 - p_x\right)^2 + \left(y_2 - p_y\right)^2}$$

Such statistical effects are used by Program (orbitdist.py) to measure the distance gap.

The square root method should be used to boost performance. We just need to consider alternative factors, up to the square root of n, instead of attempting all variables up to n − 1. In Program the sort feature (moreeff.py) reduces the number of different factors.

```python
import math
print('this floor and ceiling value of 3.56
are:'+str(math.ceil(23.56))+','+str(math.floor(3.56)))
x = 10
y = -15
print('The value of x after copying the sign from y is:
'+str(math.copysign(x, y)))
print('Absolute value of -96 and 56 are: '+str(math.
fabs(-96))+', '+str(math.fabs(56)))
my_list = [12, 4.25, 89, 3.02, -65.23, -7.2, 6.3]
print('Sum of the element of the list:'+str(math.
fsum(my_list)))
print('The GCD of 24 and 56:'+str(math.gcd(24, 56)))
x = float('nan')
if math.isnan(x):
   print('It is not a number')
x = float('inf')
y = 45
if math.isinf(x):
   print('It is Infinity')
print(math.isfinite(x))
print(math.isfinite(y))
```

Output

```
This Floor and Ceiling value of 3.56 are: 24, 23
The value of x after copying the sign from y is: -10.0
Absolute value of -96 and 56 are: 96.0, 56.0
Sum of the elements of the list: 42.13999999999999
The GCD of 24 and 56 : 8
It is not a number
It is Infinity
False
True
```

6.18 Conclusion

Python plays a very important role in minimizing the length of a program and also increases program readability, efficiency and complexity using functions. With the help of the divide and conquer approach, a large program is divided into smaller modules or functions. It is seen that functions can take as many parameters as needed and can be implemented in such a way that they perform a specific task. They can also return many values which can be used later. Functions make our code reusable and also save time as the same functions can be imported in any other program to perform that same task. Also with the recursive category of the functions, we can reduce the length of code by using recursive calls, though recursive functions are less efficient in space and time complexity as compared to normal functions.

7

Lists

7.1 Introduction to Lists

A variable can contain only one value most of the time. In certain instances, though, more can be done. Find a woodland environment in which individuals of diverse communities reside. Several of them are carnivorous, like lions, tigers and cheetahs; others such as monkeys, elephants and buffalos are herbivorous. To computationally describe the environment, you must create several variables, which may not be very helpful, to represent each species that belongs to a specific family. You should also use this variable name to integrate all the animals in the individual animal family into one number. You can see the list as a container of many objects. An object is any unit or attribute within a set. A single variable is allocated to all objects in a set [47].

Lists prohibit each object from providing a single variable that is less powerful and more vulnerable to errors when you try to run those operations on certain items. Easy lists or lists with different values may be nested lists. Listing is one of Python's most versatile storage structures since it can add, uninstall and modify attributes.

7.2 Creating Lists

Square brackets [] are used to create lists that contain comma-separated objects [48].

DOI: 10.1201/9781003202035-7

The list creation syntax is:

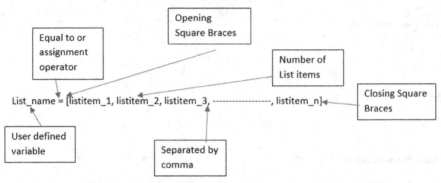

```
1. >>> country = ["England", "Russia", "Canada", "Italy",
"Spain"]
2. >>> country
["England", "Russia", "Canada", "Italy", "Spain"]
```

In 1, each object of the list is a string. The list variable contents are shown when the variable name 2 is executed. The performance is exactly the same as the list that you generated before you print the list. Without objects, you can build a blank list. The syntax is:

$$list\_name = [\ ]$$

For example,

```
1. >>> number_list = [5, 5, 6, 8, 1, 10, 11, 15]
2. >>> mixed_list = ['cat', 90.23, 70, [8, 2, 7, 4]]
3. >>> type(mixed_list)
<class 'list'>
4. >>> empty_list = []
5. >>> empty_list []
6. >>> type(empty_list) <class 'list'>
```

In mixed list 2, items are a combination of line, float, integer sort, and a list itself, while number list 1 includes objects of the same sort. By passing the name of the variable into the type function, you can evaluate a mixed list of three variable sorts. The array category is referred to in Python. As shown in lines 4–5, the empty list can be generated by list sort 6 vector blank list.

7.3 Fundamental List Operations

Lists can also be matched to Python's + sign, and the * operator is used repeatedly to construct a sequence of things in the list. For example:

```
1. >>> list_1 = [1, 2, 3, 4]
2. >>> list_2 = [5, 6, 7, 8
3. >>> list_1 + list_2
[1, 2, 3, 4, 5, 7, 8]
4. >>> list_1 * 3
[1, 2, 3, 4, 1, 2, 3, 4, 1, 2, 3,4]
5. >>> list_1 = list_2
False
```

There are generated in lines 1–2 two lists of numbers as products. A new array is added to list 1 and list 2. The elements from both lists are included in the single lists. The multiplication operator can be found on the chart. This loops the objects you assign a multitude of times and repeats them three times in list 1 and 4. List 1 and list 2 are related to operator = = 5 and the reply is Boolean negative, since the two lists are different.

By using and not by entering, you can check that the object is in the list. Boolean expression control operators or none. For instance:

```
1. >>> list_items = [1,2,3,4]
2. >>> 3 in list_items
True
3. >>> 7 in list_items
False
```

When an object is mentioned, the True 2 results in the operator but may respond False in value of Boolean 3.

7.3.1 List () Functions

For list creation, the built-in list () functionality is used [49]. The list () function syntax is:

$$list \ ([sequence])$$

where a number, tuple or list may be the sequence. When not stated as the optional number, a blank list is generated. For example:

```
1. >>> quote = "What are you doing?"
2. >>> string_to_list = list(quote)
3. >>> string_to_list
['W', 'h', 'a', 't', '', 'a', 'r', 'e''y', 'o', 'u', '',
'd', 'o', 'i', 'n', 'g', '?']
4. >>> friends = ["c", "1", "u", "b"]
5. >>> friends + quote
Traceback (most recent call last):
File "<stdin>", line 1 = in <module>
TypeError: can only concatenate list (not "str") to list
6. >>> friends + list(quote)
['c', '1', 'u', 'b', 'W', 'h', 'a', 't', '', 'a', 'r',
'e', 'y', 'o', 'u', '', 'd', 'o', 'i', 'n', 'g', '?']
```

An array conversion with the list() function 2 transforms the string reference quotation 1. Now, let's see how you can mix a string with the number 5. The show is a TypeError except that only a tuple (not 'str') can be concatenated. This implies that a set of string types cannot be merged. You need to translate the value to the list strings with Python built-in list() function 6 to concatenate the string with the number.

7.4 Slicing and Indexing in Lists

By indexing an ordered sequence of items the object in a list may also be called individually [50]. The concept is known as the "index inside the bracket." Lists use square brackets [], the first item on index 0, second item on index 1, etc., to view individual objects. The square bracket index shows the number to be obtained.

To get an object in a list, the syntax is:

```
list_name [index]
```

where the index still has to be an integer and the entity chosen should be specified.

The superstore list is shown below for index analysis.

cardstore	"metrocard"	"shopping card"	"credit card"	"R card"	"S card"
	[0]	[1]	[2]	[3]	[4]

```
1. >>> cardstore = ["metrocard", "shopping card", "credit
card", "R card", "S card"]
2. >>> cardstore [0]
'metrocard'
3. >>> cardstore [1]
' shopping card '
4. >>> cardstore[2]
' credit card '
5. >>> cardstore[3]
'R card'
6. >>> cardstore [4]
'S card'
7. >>> cardstore [9]
Traceback (most recent call last):
File "<stdin>", line 1, in <module>
IndexError: list index out of range
```

There are five products in the grocery catalog. Using parentheses immediately after list-name list with a zero 2 index value for printing the first element of the list. This list of supermarkets has index numbers from 0 to 4 2–6. The result is an error of "Index Bug: list index out of control if the index value is more than the number of the elements in column 7 [51]."

You can also view objects from the collection of negative, in addition to positive, index numbers.

Reference numbers are from the end of the row, starting at −1 and counting backward. If you have a long list and wish to search for an object at the end of a list, negative indexing is helpful.

The negative index breakdown is shown below with the same superior list.

cardstore	"metrocard"	"shopping card"	"credit card"	"R card"	"S card"
	[0]	[1]	[2]	[3]	[4]

```
1. >>> superstore [-3]
'credit card'
```

You should use a negative numerical value for the printer of the item 'credit card', as in 1.

7.4.1 Modifying List Items

Lists are mutable since once you have generated a list, the elements can be updated [52]. You can change a list by exchanging an old item with a new item and without adding a new attribute to the list. For instance:

```
1. >>> wish = ["Good morning", "Good afternoon", "good
night"]
2. >>> wish[0] = "Good evening"
3. >>> wish
["Good evening", "Good afternoon", "good night"]
4. >>> wish[2] = "Mid night"
5. >>> wish
["Good evening", "Good afternoon", "Mid night"]
6. >>> wish[-1] = "Mid noon"
7. >>> wish
["Good evening", "Good afternoon", "Mid noon"]
```

The string object at index 0 can be changed from "Good morning" to "Good evening." The list of items will be distinct (line 3) now when you showcase wish. The item on index 2 will be updated to "good night" from "Mid night". You may also use a negative index 1–6 to adjust the item's value, which fits the optimistic index number 2. Now "Mid night" is substituted for "Mid noon" 7.

If an existing list attribute is allocated to a new addition, a new copy is not rendered from lists by the transfer (=). Rather, reference refers to the same reference list with the variable names. For example:

```
1. >>> zoo = ["Lion", "Tiger". "Zebra"]
2. >>> forest = zoo
3. >>> type(zoo)
<class 'list'>
4. >>> type(forest)
<class 'list'>
5. >>> forest
['Lion', 'Tiger', 'Zebra']
6. >>> zoo[0] = "Fox"
7. >>> zoo
['Fox', 'Tiger', 'Zebra']
8. >>> forest
['Fox', 'Tiger', 'Zebra']
9. >>> forest[1] = "Deer"
10. >>> forest
['Fox', 'Deer', 'Zebra']
11. >>> zoo
['Fox', 'Deer', 'Zebra']
```

The above-noted code shows that a newer copy would not generate an original list element for an old list object. In Python, it is permissible to slide lists, where a section of the list may be retrieved by defining the index range together with the colon (:) operator which seems to be a number.

The list cutting syntax is:

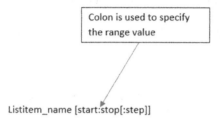

Listitem_name [start:stop[:step]]

where both start and end values (positive or negative values) are integer numbers. List slice provides a portion of the set, which contains the value of the start index but removes the value from the value at the beginning index to avoid the index value. Phase stipulates the slice raises the value and is optional. The positive and negative overview of the index is given below for a list of vegetables [53].

Cabbage	Lady finger	Tomato	Chilli	Garlic
0	1	2	3	4
−5	−4	−3	−2	−1

```
1. >>> vegetable = ["Cabbage", "Lady finger", "Tomato",
"Chilli", "Garlic"]
2. >>> vegetable [1:3]
['Lady finger', ' Tamato ']
3. >>> vegetable [:3]
['Cabbage', 'Lady finger', 'Tomato']
4. >>> vegetable [2:]
['Tomato', 'Chilli', 'Garlic']
5. >>> vegetable [1:4:2]
['Lady finger', 'Chilli']
6. >>> vegetable [:]
['Cabbage', 'Lady finger', 'Tomato', 'Chilli', 'Garlic']
7. >>> vegetable [::2]
['Cabbage', 'Tomato', 'Garlic']
8. >>> vegetable [::-1]
['Garlic','Chilli','Tomato','Lady finger','Cabbage']
9. >>> vegetable [- 3:-1]
['Tomato', 'Chilli']
```

Both items on the vegetable list from 1 to 3, except 3. Both items on the veg- etable list are slimmed 2. To reach the starting objects, the index value of zero is not required. You should miss the value for the start index and then men- tion 3 for the stop index. Likewise, if the last stop objects are to be reached, no stopping value must be indicated and the start index value 4 must be specified instead. The list of accessed documents stretches up to the last item if they stop index value is missed. If you miss both the beginning and end parameter estimates 6 and just mention the colon operators in the parenthe- ses, you can show the whole item of the list. After the second colon, the num- ber tells Python you want to pick your slice rise. For example, Python sets this raise to 1, but after the second colon, this amount lets you decide what it is. When the double colon is shown at 7, it is not the start index value and no stop index value, and the objects are jumped by two steps. -- second item in the list will be derived from a zero index value. All items in a list may be reversed, with a double column and an index value of 8 specified. For start and end index values 9 negative index values may also be used.

7.5 Built-In Functions Used in Lists

A collection can be provided as a built-in function (Table 7.1). There are sev- eral built-in functions [53.]

TABLE 7.1

Built-in Functions in Lists

Built-in Functions	Description
len()	The len() function returns the number of items in a list.
sum()	The sum() function returns the sum of the numbers in the list.
any()	The any() function returns True if any of the Boolean values in the list are True.
all()	The all() function returns True if all the Boolean values in the list are True, else it returns False.
sorted()	The sorted() function returns a modified copy of the list while leaving the original list untouched.

For example:

```
1. >>> rivers = ['satluj', 'kavari', 'nokia', 'jehlem',
       'godavri']
2. >>> len(rivers)
```

```
5
3. >>> numbers = [1, 2, 3, 4, 5]
4. >>> sum(numbers)
15
5. >>> max(numbers)
5
6. >>> min(numbers)
1
7. >>> any([1, 1, 0, 0, 1, 0])
True
8. >>> all([1, 1, 1, 1])
True
9. >>> rivers_sorted_new = sorted(rivers)
10. >>> rivers_sorted_new
['godavri', 'jehlem', 'kavari', 'nokia', 'satluj']
```

The len() function 2 is used to determine the number of objects in the list river. The add-on of all the numbers in list 4 results with the Sum() function. Using max() 5 and min() functions 6, higher and lower numbers in a set will be returned. If one of the things in the list has 1, the Boolean Truth value 7 is returned by every() function. All functions return True, while all of the elements in the array are 1 the False Boolean attribute is 8. Without changing the original list which is allocated to a new list rivers 9, sorted() returns the sorted list of items. For string types of items, they are ordered according to their ASCII values.

7.6 List Methods

When you add or delete items, the list dimensions change dynamically and you do not have to handle them by yourself [54]. A list of all methods (Table 7.2) associated with the list will be provided to dir() through the list function.

TABLE 7.2

Different Types of List Method

Dictionary Methods	Syntax	Description
clear()	dictionary_name. clear()	The clear() method excludes from the dictionary any key:value pairs.
fromkeys()	dictionary_name. fromkeys(seq [, value])	From the sequence of elements supplied, the fromkeys() method generates a new dictionary with the user value.

(Continued)

TABLE 7.2 (CONTINUED)

Different Types of List Method

Dictionary Methods	Syntax	Description
get()	dictionary_name. get(key [, default])	The get() method returns the value of the key in the dictionary defined. If the key is not available, the default value is returned. If no default is issued, None is the default, so KeyError is never raised by that process.
items()	dictionary_name. items()	The items() method gives a new view of the key and value pairs of a dictionary as tuples.
keys()	dictionary_name. keys()	A new view consisting of all keys in the dictionary is returned using key() procedure.
pop()	dictionary_name. pop(key[, default])	The pop() method extracts the key and returns its meaning from the dictionary. If the key is missing, the default value will be returned. The outcome would be KeyError if the default is not given and the key is not in the dictionary.
popitem()	dictionary_name. popitem()	The popitem() procedure eliminates a tuple pair arbitrary to the dictionary (key, value). If the dictionary is empty, the KeyError results in calling popitem().
setdefault()	dictionary_name. setdefault (key[, default])	A value for the key in the dictionary is returned by a setdefault() method. If the key is not available, insert the key and return the default value to the dictionary. The default is None if the key is present, so that a KeyError is never raised.
update()	dictionary_name. update([other])	The update() method uses the key:value pairs of other dictionary objects to update the dictionary and returns none.
values()	dictionary_name. values()	A new view consisting of all the dictionary values is returned by the value() process.

For example:

```
1. >>> dir( list)
['__add__', '__class__', '__contains__', '__delattr__',
'__delitem__',. '__dir__' '__doc__',
'__eq__', '__format__', '__ge__', '__getattribute__', '
__getitem__', '__gt__', '__hash__'
'__iadd__', '__imul__', '__init__', '__init_
subclass__','__iter__', '__le__', '__len__',
'__It__', '__mul__', '__ne__', '__new__', '__reduce__',
'__reduce_ex__', '__repr__',
```

```
'__reversed__', '__rmul__', '__setattr__', ' __setitem__'
'__sizeof__', '__str__',
'__subclasshook__', 'append', 'clear', 'copy', 'count',
'extend', 'index', 'insert', 'pop',
'remove',
"reverse', 'sort']
```

```
1.>>> town = ["bombay","kolkata","paris","new york","pune
   ","mumbai","Rohtak"]
2.>>> town.count("new york")
1
3.>>> town.index("pune")
4
4.>>> town.reverse()
5.>>> town
['Rohtak', 'mumbai', 'pune', 'new york', 'paris',
   'kolkata', 'bombay']
6.>>> town.append("sonipat")
7.>>> town
['Rohtak', 'mumbai', 'pune', 'new york', 'paris',
   'kolkata', 'bombay', 'sonipat']
8.>>> town.sort()
9.>>> town
['Rohtak', 'bombay', 'kolkata', 'mumbai', 'new york',
   'paris', 'pune', 'sonipat']
10.>>> town.sort()
11.>>> town
['Rohtak', 'bombay', 'kolkata', 'mumbai', 'new york',
   'paris', 'pune', 'sonipat']
12.>>> town.pop()
'sonipat'
13.>>> town
['Rohtak', 'bombay', 'kolkata', 'mumbai', 'new york',
'paris', 'pune']
14.>>> more_town = ["mogga","amritsar"]
15.>>> town.extend(more_town)
16.>>> town
['Rohtak', 'bombay', 'kolkata', 'mumbai', 'new york',
   'paris', 'pune', 'mogga', 'amritsar']
17.>>> town.remove("mogga")
18.>>> town
['Rohtak', 'bombay', 'kolkata', 'mumbai', 'new york',
   'paris', 'pune', 'amritsar']
```

Different list operations are performed by list methods 1–18.

7.6.1 Populating Lists Items

One of the common ways to compile lists is to start with a blank list [] and then to add objects to it by the functions append() or extend() [55]. For example:

```
1. >>> countries = []
2. >>> countries.append("UAE")
3. >>> countries.append("USA")
4. >>> countries.append("UK")
5. >>> countries
['UAE', 'USA', 'UK']
```

This creates a vacant list of countries 1 and begins to add things to the list of countries

Section 2–4 Append() and ultimately lists five countries' objects shown.

7.6.2 List Traversing

Each object is iterated using a single loop in a list.

Example: Design a program for traversing lists using the for loop

```
fast_food = ["chowmin", "sandwich", "burger", "fries"]
for each_food_item in fast_food:
    print(f"I like to eat {each_food_item}")
for each_food_item in ["waffles", "sandwich", "burger",
"fries"]:
    print(f"I like to eat {each_food_item}")
```

Output

```
I like to eat chowmin
I like to eat sandwich
I like to eat burger
I like to eat fries
I like to eat chowmin
I like to eat sandwich
I like to eat burger
I like to eat fries
```

It is simple to loop around objects in a list with this statement. There is a list variable 1 and a list variable in for loop 2. The array may be defined explicitly in loop 4 instead of defining an array attribute. By using the range() and len() function you can obtain the index value of each object in the set.

7.6.3 Nested Lists

In another list, a list of images is named. Nested lists in Python can be operated by storage lists in other list components [56]. An immobilized list with a loop can be shifted.

The nesting list syntax is:

```
                              Opening Braces
              nested_list_name = [[item_1, item_2, item_3],  Comma seperated list-item
   user defined list name       [item_4, item_5, item_6],
                                [item_7, item_8, item_9]]
                              Nested List-Items    Closing Braces
```

Each list is divided by a comma inside another list. For instance:

```
1. >>> asia = counting = [["one", "two", "three"],
["four", "five", "six"],
["seven", "eight", "nine"]]
2. >>> counting[0]
["one", "two", "three"]
3. >>> counting[0][1]
'two'
4. >>> counting[1][2] = "ten"
5. >>> counting
[["one", "two", "three"],
["four", "five", "six"],
["seven", "eight", "nine"]]
```

You can join an object within a list which is itself in an individual list by channeling three sets of square parentheses combined ["one", "two", "three"], ["four", "five", "six"], ["seven", "eight", "nine"]. For example, when you have three lists (line 1) with the 3 by 3 matrix in the counting variable in the list above. If the objects in the first column are to be shown then state the column variable followed by the list index you need to navigate in brackets, such as counting[0] (line 2). If you want access to "two" in the list, you must state the list's index in the section, followed by the section's index, such as counting[0][1] (line 3). In the list, you can also change the contents. In line 4, for instance, the code is used to override "six" by "ten."

7.7 Del Statement

You may delete an object based on its index instead of its value from a list. The distinction between method del and pop() is that del does not return a value, while pop() displays the result. You may also use del to delete the list slices or clean the entire list.

```
1. >>> b = [9, 56, 45.12, 25, 136, 213]
2. >>> del b[0]
3. >>> b
[56, 45.12, 25, 136, 213]
4. >>> del b[2:4]
5. >>> b
[9, 56, 136, 213]
6. >>> del b[:]
7. >>> b
[]
```

This deletes an object with a 0 index value 2. Now the original list has the number of things 3. The objects with an index value of 2 to 4 are excluded with list 4, but not from an index value of 4. Only if the colon operator is set without the index value, will all items on the list be removed 6–7.

7.8 List Operations

List operations are necessary for searching or sorting different types of lists [57].

7.8.1 Searching Problem

A common activity is to search for the list of a particular element. Two main approaches are explored: linear search and binary search.

7.8.1.1 Linear Search

A feature called a place returns the first position of occurrence for the item specified in the list of Program (linearsearch.py) below. If the item does not occur, the function returns zero.

```
def locate(lst, seek):
    '''
Returns the index of element seek in list lst,
```

```
if seek is present in lst.
Returns None if seek is not an element of lst.
lst is the lst in which to search.
seek is the element to find.'''
  for i in range(len(lst)):
    if lst[i] == seek:
      return i # Return position immediately
    return None # Element not found

def format(i):
  if i > 9999:
    print("****") # Too big!
  else:
    print("%4d" % i)

def show(lst):
  ''' Prints the contents of list lst '''
  for item in lst:
    print("%4d" % item, end='') # Print element right
    justifies in 4 spaces
  print() # Print newline

def draw_arrow(value, n):
  print(("%" + str(n) + "s") % " ^ ")
  print(("%" + str(n) + "s") % " | ")
  print(("%" + str(n) + "s%i") % (" +-- ", value))

def display(lst, value):
  show(lst) # Print contents of the list
  position = locate(lst, value)
  if position != None:
    position = 4*position + 7; # Compute spacing for arrow
    draw_arrow(value, position)
  else:
    print("(", value, " not in list)", sep='')
  print()

def main():
  a = [100, 44, 2, 80, 5, 13, 11, 2, 110]
  display(a, 13)
  display(a, 2)
  display(a, 7)
  display(a, 100)
  display(a, 110)

main()
```

Output

```
100   44   2  80   5  13  11    2 110
                        ^
                        |
                        +-- 13

100   44   2  80   5  13  11    2 110
                  ^
                  |
                  +-- 2

100   44   2  80   5  13  11    2 110
(7 not in list)

100   44   2  80   5  13  11    2 110
  ^
  |
  +-- 100

100   44   2  80   5  13  11    2 110
                             ^
                             |
                             +-- 110
```

This program has the main function; all other functions merely result in an informative view of the effects of the position. If location detects the match, the method returns the corresponding element's place immediately; otherwise, if the item retrieved cannot be located after analysis of all the elements of the collection, it returns zero. Zero here suggests that a correct response could not be given to the question.

The customer code, the display function in this example, must ensure that the results are zero before attempting to index the results into a list.

The method of search conducted by a locate is called a linear search, since the beginning of the list to the end of each item is followed by a straight line. The linear analysis is shown in Figure 7.1.

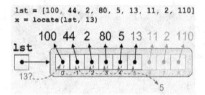

FIGURE 7.1
Linear Search.

7.8.1.2 Binary Search

With relatively short items, linear searching is appropriate, but it requires time to analyze any object on a broad list. Binary search is an alternative to linear search. A collection must be ordered to execute a binary scan. The binary search uses a sophisticated yet easy technique to use the sorted structure of the list to rapidly home in on the element:

1. Return None if the list is zero.
2. In the center of the list, test the item.

If you are searching for the function, return its location. Fill in the first half of the variable, if the center of the item is bigger than the category you are searching for. Find a binary; check in the second if the intermediate components are narrower than the item you are looking for, as half the list is. This framework is like finding a telephone number in the book:

- Open the book's center. In one of the two pages you see the telephone number as the names of each display.
- If not, the investigation applies alphabetically to the left half, and the last name of the person less the names on the recognizable paragraphs, then the query appears to apply otherwise to the right half of the open book.
- If the name of the individual should be on one of the two visible pages but is not present, discontinue searching with failure.

As seen in Program (binarysearch.py), a Python method can be used for the binary search algorithm:

```
def binary_search(lst, seek):
  first = 0 # Initialize the first position in list
  last = len(lst) - 1 # Initialize the last position in list
  while first <= last:
    mid = first + (last - first + 1)//2 # Note: Integer
    division
    if lst[mid] == seek:
      return mid # Found it
    elif lst[mid] > seek:
      last = mid - 1 # continue with 1st half
    else: # v[mid] < seek
      first = mid + 1 # continue with 2nd half
  return None # Not there

def show(lst):
  for item in lst:
```

```
    print("%4d" % item, end='') # Print element right
        justifies in 4 spaces
  print() # Print newline

def draw_arrow(value, n):
  print(("%" + str(n) + "s") % " ^ ")
  print(("%" + str(n) + "s") % " | ")
  print(("%" + str(n) + "s%i") % (" +-- ", value))

def display(lst, value):
  show(lst) # Print contents of the list
  position = binary_search(lst, value)
  if position != None:
    position = 4*position + 7; # Compute spacing for arrow
    draw_arrow(value, position)
  else:
    print("(", value, " not in list)", sep='')
  print()

def main():
  a = [2, 5, 11, 13, 44, 80, 100, 110]
  display(a, 13)
  display(a, 2)
  display(a, 7)
  display(a, 100)
  display(a, 110)

main()
```

Output

```
     2   5  11  13  44  80 100 110
                  ^
                  |
                +-- 13

     2   5  11  13  44  80 100 110
         ^
         |
       +-- 2

     2   5  11  13  44  80 100 110
   (7 not in list)

     2   5  11  13  44  80 100 110
                              ^
                              |
                            +-- 100

     2   5  11  13  44  80 100 110
                                  ^
                                  |
                                +-- 110
```

The working of binary search is shown in Figure 7.2. It is more difficult to apply the binaries search algorithm Easier than a linear search. Normally quicker and easier, but more intelligent, techniques use smart tricks to try to influence the data format (as binary search does) which typically outweigh simpler, easier-to-code algorithms that handle data structures that probably contain massive amounts of data – if we want to locate an item in a sorted list for a fair comparison between linear and binary searches. The ordering of the list is crucial for a binary search, but it can also enable linear searching. Revised Algorithm of linear method looking for structured lists

```
# This version requires list 1st to be sorted in #
ascending order.
def linear_search(1st, seek):
        i = 0        # Start at beginning
n = len(1st)# Length of list
    while i < n and 1st[i] <= seek:
    if 1st [i] == seek:
                            return i    # Return position
                            immediately
                    return None # Element not found
```

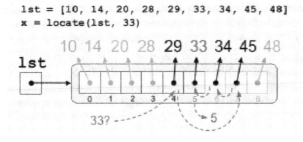

```
1st = [10, 14, 20, 28, 29, 33, 34, 45, 48]
x = locate(1st, 33)
```

FIGURE 7.2
Binary search.

It must be remembered that the loop stops after all the items have been checked, as in the initial iteration of linear search, although this variant finishes faster when an item greater than that required is located in it. Because the list is sorted, the search does not proceed until an element that is greater than the value sought is identified in the index; the search cannot start after a greater item in the sorted list.

Suppose there are n elements in a list to search. The worst aspect is that the process of linear search results of iterations with an item that is greater than every other entity in the list now. When an item is less than those the technology automatically returns in the list without consideration of the other feature items into consideration for the ideal case. Therefore, a typical linear search needs a 2n comparison of loop iterations before the loop termination and the feature returns, varying from 1 to n.

Think binary search now. The lists left to remember are half the original size after increasing compare. If the item you are looking for is not found on the first probe, the number is n^2. The next time you move into the sequence, there are decreases to n^4, then n^8, and so on. An algorithm of base 2 is a challenge in deciding how much a whole collection of items can be partitioned into half up to the point where only one element remains. For a binary search, a loop requires $\log_2 n$ iterations in the worst-case situation not to find the object searched.

How do we decide which search is best by using this analysis? Two main characteristics assess the efficiency of an algorithm:

- How long will it take to run (processor cycles)?
- How much memory is needed for it to operate?

In our scenario, the search algorithms process the list only with a few additional local variables, and they all use the same space for big lists. Here, speed makes a big difference. A binary search conducts more complex calculations each and every loop, such that any process requires more time. Linear scanning is easier (fewer loop operations), but the process might be more time-intensive than the binary search method because it is slower overall.

In logical and theoretical order, we can induce the faster algorithm. An observational evaluation is an experiment; all methods are cautiously applied, and implementation times are calculated. Analytically, the source code is evaluated to determine how many transfers the application processor has to execute to enforce the system on a given size issue.

We obtain certain observational findings in Program 7.8.1.2 (comparisonof-search.py):

```python
def sequentialSearch(alist, item):
    pos = 0
    found = False
    while pos < len(alist) and not found:
        if alist[pos] == item:
            found = True
        else:
            pos = pos + 1

    return found
def binSearch(list, target):
```

```
    list.sort()
    return binSearchHelper(list, target, 0, len(list) - 1)

def binSearchHelper(list, target, left, right):
    if left > right:
        return False

    middle = (left + right)//2
    if list[middle] == target:
        return True
    elif list[middle] > target:
        return binSearchHelper(list, target, left, middle - 1)
    else:
        return binSearchHelper(list, target, middle + 1,
            right)

import random
import time
list_sizes = [9,99,999,9999,99999,999999]
for size in list_sizes:
    list = []
    for x in range(size):
        list.append(random.randint(1,9999999))

    sequential_search_start_time = time.time()
    sequentialSearch(list,-1)
    sequential_search_end_time = time.time()
    print("Time taken by linear search is = ",(sequential_
    search_end_time-sequential_search_start_time))

    binary_search_start_time = time.time()
    binSearch(list,-1)
    binary_search_end_time = time.time()
    print("Time taken by binary search is =
    ",(binary_search_end_time-binary_search_start_time))

    print("\n")
```

Output

```
Time taken by linear search is = 0.0
Time taken by binary search is = 0.0

Time taken by linear search is = 0.0
Time taken by binary search is = 0.0

Time taken by linear search is = 0.0
Time taken by binary search is = 0.0
```

```
Time taken by linear search is = 0.0029993057250976562
Time taken by binary search is = 0.0029973983764648438

Time taken by linear search is = 0.028982162475585938
Time taken by binary search is = 0.043975114822387695

Time taken by linear search is = 0.29885029792785645
Time taken by binary search is = 0.6685905456542969
```

In this program the test search method checks through all objects in the array with a first ordered linear check and then a binary.

The organized linear search experiments were carried out in less than one-fifth of a second with an exact search of 72 seconds. A binary quest is almost 400 times quicker than linear search orders!

The findings for various size lists are provided in Table 7.3. A binary quest is, empirically speaking, much easier than a linear search.

We may also determine whether the algorithm is best evaluated with each function in its source code using an analytical method. This is the time to run any arithmetical operation, task, logical comparison, and list access. They must presume that each of these operations involves a single unit of processing time, although this statement is not purely valid. Given that both search algorithms are analyzed using the same rules, relative results will be fairly accurate for comparison purposes.

A linear search is the first step. On average, we find that the loop generates n^2 iterations for a size n array. It is initialized just once in a linear-search query. Any other operation affecting the loop happens 2^n times except for return comments. It will be retrieved, and during every request, only one return will be performed. An overview of linear analysis is shown in Table 7.4.

The iterations of the 2n loop are focused in the meantime on finding an element. During a specified request, the method must perform precisely one of the two statements of return.

TABLE 7.3

Comparison of Linear Search and Binary Search

List Size	Linear Search	Binary Search
9	0.0	0.0
99	0.0	0.0
999	0.0	0.0
9999	0.015630245208740234	0.006513118743896484
99999	0.06902360916137695	0.0896601676940918
999999	0.717473030090332	1.2188801765441895

TABLE 7.4

Analysis of Linear Search Algorithm

Action	Operation(s)	Operation Count	Times Executed	Total Cost
pos = 0	=	1	1	1
found = False	=	1	1	1
while pos < len(alist)	While < len and	4	n/2	2n
and not found:	If [] == :	4	n/2	2n
if alist[pos] == item:	=	1	1	1
found = True	= +	2	n/2	n
else:	return	1	1	1/2
pos = pos + 1				
return found				

Note: Total time units $3n+4$

TABLE 7.5

Binary Search Algorithm (BSA) Analysis

Action	Operation(s)	Operation Count	Times Executed	Total Cost
def binSearchHelper(list, target, left, right):	()	1	1	1
if left > right:	>	1	1	1
return false	return	1	1	1
middle = (left + right)//2	=, +, //	3	$\log_2 n$	$3\log_2 n$
if list[middle] == target:	[], ==	2	$\log_2 n$	$2\log_2 n$
return True	return	1	1	1
elif list[middle] > target:	[], >	2	$\log_2 n$	$2\log_2 n$
return binSearchHelper(list, target, left, middle - 1)	(), -	2	$\frac{1}{2}\log_2 n$	$\log_2 n$
else:		0		0
return binSearchHelper(list, target, middle + 1, right)	return, (), +	3	$\log_2 n$	$3\log_2 n$

Note: Total time units $10\log_2 n+6$

Table 7.4 displays the term linear search time for a basic mathematics linear equation:

$$f(n) = 3n + 4.$$

Next, is a binary search. In the worst situation, we have calculated that the loop is $\log_{2n}$ times while the collection includes n elements in a binary search. The two start-ups are made once per call before the loop. Many loop activities occur $\log_{2n}$ times, except that only one return state can be performed per request and that only a single route can be picked per loop iteration for the if / elif / else argument. Table 7.3 displays the complete binary search analysis. The two functions are 3n + 4 and $12 \log_{2n} + 6$.

At the end, the binary search is quick, even for big lists.

```python
def binSearchHelper(list, target, left, right):
    if left > right:
        return False

    middle = (left + right)//2
    if list[middle] == target:
        return True
    elif list[middle] > target:
        return binSearchHelper(list, target, left, middle - 1)
    else:
        return binSearchHelper(list, target, middle + 1, right)
```

Any time method performs `elif` or other assertions across the loop; the real cost of a single one is compensated.

7.8.2 Sorting

It is common to sort the elements in a list into a particular order. For example, the list of integers can be arranged in ascending order (from smaller to bigger). Lexicogram (usually known as alphabetical) order may be used to group a collection of strings. There are several sorting algorithms, and some of them function much better than others. We should call a sorting algorithm that can be applied fairly quickly.

It is very easy to apply the sorting algorithm and quickly comprehend how it operates. If A is a variable and I is a collection index, the type selection works as follows:

1. "Set n = list longitudinal A.
2. Set I = 0.
3. Review all A[idx+1] elements, idx < idx+1 < n. (If one of those elyesements is less than A[i], then compare A[idx] with the lesser of such elements. This means literally finding all of the elements in the sequence from index I to the final one. This ensures that all elements are higher or equal after position I.)

4. If *i* is less than *n*−1, set *i* equal to *i*+1 and go to Step 2
5. Set to i+1 if I is less than n−1 and go to Step 2.
6. Sorted Done (list)"

The "go to step 2" order in step 4 is a loop. The algorithm finishes with a sorted list when the I in Step 3 equals n.

The above definition can be interpreted as follows in Python:

```
def sort(list):

    for iter_num in range(len(list)-1,0,-1):
        for idx in range(iter_num):
            if list[idx]>list[idx+1]:
                temp = list[idx]
                list[idx] = list[idx+1]
                list[idx+1] = temp

list = [65,45,75,84,15,2,4,56,12,20]
sort(list)
print(list)
```

Output

```
[2, 4, 12, 15, 20, 45, 56, 65, 75, 84]
```

We also added a new small variable to evaluate if each of the elements is less than A[idx+1]. The limited goal is to pursue the location of the smallest item that has been discovered up to now. We will start with the set smaller than 1, since in that position we want to find anything less than the item.

7.8.2.1 Selection Sort

The selection sort algorithm sorts an array by repeatedly finding the minimum element (in ascending order) from the unsorted part and putting it at the beginning. The algorithm maintains two subarrays in a given array:

1. The subarray which is already sorted;
2. The remaining subarray which is unsorted.

In every iteration of selection sort, the minimum element (in ascending order) from the unsorted subarray is picked and moved to the sorted subarray (Figure 7.3).

It may be daunting to create a complete solution on the first attempt. To begin with, write the first iteration code, to define the smallest item in the

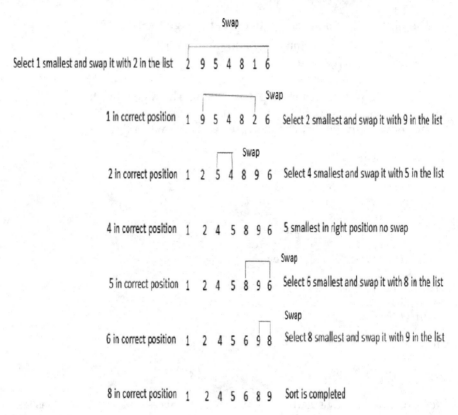

FIGURE 7.3
Selection sort.

list, swap it for the first item, then see what else will be, then the third. You should write a loop that shows that you are intuitive about these things.

The following solution can be described:

```python
def selection_sort(input_list):

    for idx in range(len(input_list)):
        min_idx = idx
        for j in range(idx+1, len(input_list)):
            if input_list[min_idx] > input_list[j]:
                min_idx = j

        input_list[idx], input_list[min_idx] = input_list[min_idx], input_list[idx]

l = [2, 9, 5, 4, 8, 1, 6]
selection_sort(l)
print(l)
```

Output

```
[1, 2, 4, 5, 6, 8, 9]
```

7.8.2.2 Merge Sort

Merge sort is an algorithm for dividing and conquering. The input panel is separated in two halves, which are then called up and fused. For combining two halves, the merge() feature is used. A main mechanism is the merge (arr, l,m,r), which means Ar[s] and Ar[m+1..r] have been sorted and the two sorted sub-arrays have been combined into a single one. Figure 7.4 displays the entire merge phase in order to create an example array {38, 27, 43, 3, 9, 82, 10}. If we look at the diagram in more detail, we can see that the list is separated into two halves until its size is 1. After size 1, the merge is started and arrays are fused back before the whole array is combined.

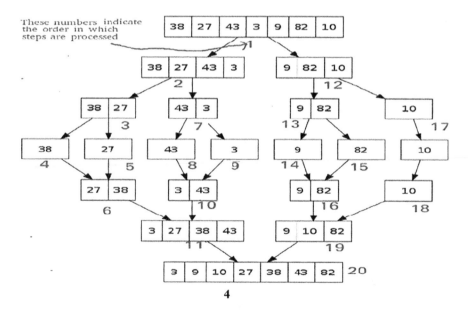

FIGURE 7.4
Merge sort.

```
def merge_sort(unsorted_list):
    if len(unsorted_list) <= 1:
        return unsorted_list
# Find the middle point and devide it
    middle = len(unsorted_list) // 2
    left_list = unsorted_list[:middle]
    right_list = unsorted_list[middle:]

    left_list = merge_sort(left_list)
    right_list = merge_sort(right_list)
    return list(merge(left_list, right_list))

# Merge the sorted halves

def merge(left_half,right_half):

    res = []
    while len(left_half) != 0 and len(right_half) != 0:
        if left_half[0] < right_half[0]:
            res.append(left_half[0])
            left_half.remove(left_half[0])
        else:
            res.append(right_half[0])
            right_half.remove(right_half[0])
    if len(left_half) == 0:
        res = res + right_half
    else:
        res = res + left_half
    return res

unsorted_list = [38, 27, 43, 3, 9, 82, 10]

print(merge_sort(unsorted_list))
```

Output

```
[3, 9, 10, 27, 38, 43, 82]
```

7.8.2.3 *Sorting Comparison*

Merge is a very real-life example of an algorithm for dividing and conquering. The two key phases of this algorithm are easily used:

- Continuously split the unordered list until you have N sublists, where each sublist has 1 component that is "unsorted" and N represents the total of items in the original sequence.

- Repeatedly combine the two sublists to generate newly ordered sublists before all items have been completely merged in a single categorized sequence.

The selection sort of variety is often very easy but also reaches a sort of bubble. If you choose between the two, it's easiest to sort only by chance. We break the input list/array into two sections with the sort sub-list of the previously sorted items and the sublist of the remaining items which make up the rest of the list. The smallest element is located first in the unsorted sublist and at the end of the sorted sublist. Thus, we catch the little unsorted part constantly and put it in the sorted sublist in order. This process continues until the list is worked out in its entirety.

7.9 Reversing of Lists

Program (listreverse.py) includes a recursive feature called `rev`, which accepts a table as a parameter and returns a new list in reverse order of all items in the original list.

```
systems = ['Windows', 'mac OS', 'Linux']
print('Original List:', systems)

# List Reverse
systems reverse()

# updated list
print("Updated List:', systems)
```

Output

```
Original List: ['Windows', 'macOS', 'Linux']
Updated List: ['Linux', 'macOS', Windows']
```

Python has a flipped basic feature taking a set parameter. Instead, the reverse function returns an iterable entity that can be used, equivalent to the range method in a loop which shows how reverse may be used to reverse the file content.

7.10 Conclusion

This chapter has covered lists in depth: the basics of list creation, indexing and slicing. We have briefly compared lists to other data structures, understood multidimensional lists, seen how to iterate lists, and combine and operate them on Python lists. A Python list is mutable. It provides various functions to add, insert, update and remove its elements.

8

Dictionaries

8.1 Introduction

The dictionary is yet another valuable form of data implemented in Python. You see your email list on your computer in the actual world. The mobile number of people you know is difficult to memorize. You store the person's name and number in the "Contacts" file. You may classify the mobile phone number based on the names of the user. The name of an individual can be interpreted as the key to his or her mobile phone, which is a key attribute. Writings in other languages, like "associative memories" or "associative arrays," are sometimes found [58].

8.2 How a Dictionary Is Created

A dictionary consists of a group of unordered keys: value pairs which indicate that the key in a dictionary is unique. Dictionaries are designed with curly braces{}, which contain a set of commas separated by key: value pairs. In addition to this, there is a column (:) dividing each key and value pair where the words to the left of the colon operator are the keys, and those to the right are the values. Dictionaries are indexed by the keys unlike lists which are indexed by a set of numbers. Here a key is called a key: value pair along with its corresponding value. Dictionary keys are case sensitive.

The syntax of dictionary formation is:

```
Dictionary-name = {key-1: value-1, key-2:
value-2,............... key-n: value-n}
```

DOI: 10.1201/9781003202035-8

For example:

```
1. >>>wild_animal = {"t": "tiger", "l":"lion", "p":
"panther", "b":"black panther",
 "e":"elephant"}
2. >>> wild_animal
{"t": "tiger", "l":"lion", "p": "panther", "b":"black
panther", "e":"elephant"}
```

In the curly braces, inserting an input-separated set of key: value pairs provides an initial core: the dictionary value pairs. That's how dictionaries on output 2 are written. The keys in wild_animal are dictionary "t", "l", "p", "b" and "e" and "tiger", "lion", "panther", "black panther" and "elephant". Each string type contains keys and values. Every data form including string, number, list or dictionary itself may be a valuable feature in the dictionary. The keys and related values in dictionaries can be of various kinds.

For example:

```
1. >>>school_dict = {"student":"english", 11,
5feet:"abcd"}
2. >>>school_dict
{"student":"english", 11, 5feet:"abcd"}
3. >>>type(school_dict)
<class 'dict'>
```

school_dict, keys and key considerations are all multiple kinds in the dictionary. It doesn't need to begin with zero for numbers as keyboards in dictionaries can be any number. The type of school_dict can be calculated by forwarding the variable name to the form () function as an argument. The dictionary type is known as a dictionary in Python. By specifying a pair of curly braces, you can create an empty dictionary key:value pairs.

The syntax is:

$$dictionary\_name = \{ \ \}$$

For example:

```
1. >>>empty_dictionary = {}
2. >>>empty_dictionary
{}
3. >>>type(empty_dictionary)
<class 'dict'>
```

As shown in line 1, empty dictionary is in dictype 3, the order of the key : value pairs doesn't really mean a thing in dictionaries. For example:

```
1. >>>wheat = {"biscuit":3, "spring roll":5, "chappti":4}
2. >>>Bajara = {"chappti":4, "biscuit":3, "spring
roll":5}
3. >>>wheat == Bajara
True
```

The wheat and Bajara dictionaries have almost the same key: value pairs in order from lines 1 to 2. When you distinguish two dictionaries, a Boolean True Value 3 results. This implies that it doesn't make a difference in dictionaries to order key: value pairs. Slicing is not permitted in dictionaries because the lists are not ordered.

8.3 Accessing and Altering Key: Value Pairs in Dictionaries

A single key: a dictionary value pair can be obtained by accessing the square brackets while using keys [59]. The key inside the square brackets reveals the address meaning pair to be obtained.

The significance of a dictionary key is in the syntax:

```
dictionary_name [key]
```

The syntax for altering the value or inserting a new key: a dictionary value pair is:

```
dictionary_name[key] = value
```

If the driver key is still in the dictionary, the word will be replaced by the new meaning. The new key: value pair is associated with the dictionary if the key is not available. For example:

			Festival		
Keys	"holi"	"teej"	"rakhi"	"Deepawali"	"lodi"
Values	5225	6225	7225	8225	9225

```
1. >>>festival = {"holi":5225, "teej":6225, "rakhi":7225,
"Deepawali":8225, "lodi":9225}
2. >>>festival["holi"] = 4225
3. >>>festival
{"holi":4225, "teej":6225, "rakhi":7225,
"Deepawali":8225, "lodi":9225}
4. >>> festival["rakhi"]
7225
5. >>>festival["makar sankarnti"] = 3225
6. >>>festival
{"holi":4225, "teej":6225, "rakhi":7225,
"Deepawali":8225, "lodi":9225, "makar sankarnti":3225}
7. >>>festival["hellofest"]
Traceback (most recent call last):
File "<stdin>", line 1, in <module>
KeyError: 'piero'
```

The dictionaries are changeable, so the assignment operator may introduce a new key: value couple or change the current key values. Original values for the keys can be updated for the Festival 1 dictionary. As seen in line 2, in the assignment operator, the value of a central Holi is modified from 5225 to 4225. The meaning for the Rakhi key is seen in line 4. A new key: value pair can be inserted by defining the spelling of the dictionary and then the bracket in the main name and assigning it a value 5. If you try to use an existing key 7, it will lead to KeyError. In membership operators, you can track the existence of a key in the dictionary. It will return a True or False Boolean value. For example:

```
1. >>>season = {"rainy":"umbrella", "summer":"AC",
"winter":"Blanket"}
2. >>>"spring" in umbrella
False
3. >>>"spring" not in umbrella
True
```

In lines 2 and 3 in the dictionary of clothes the existence of the key spring is tested.

8.3.1 The dict () Function

The built-in dict() functionality is used for dictionary development [60]. An alternative keyword argument can be used: dict() syntax

dict([**kwarg])

`Dict()` returns the entire original dictionary and a collection of keyword arguments which may be null from the optional keyword declaration. If no keyword argument is made, a dictionary is created instead. The keyword arguments and `Kwarg = value` are added as the keyword parameters for the dictionary if the argument is given.

For example:

```
1. >>>digits = dict(four=4, five=5, six=6)
2. >>>digits
{'four': 4, 'five': 5, 'six': 6}
```

For the digits of dictionary numbers 1–2 the keyword arguments in the form `kwarg= value` are translated into `key: value` pairs.

The `dict()` function syntax for using iterables is:

$$dict(iterable[, **kwarg])$$

An iterable with exactly two items can be defined, as multiple, key and the value of the function `dict()`. For example

```
1. >>>dict([('sample', 6666), ('wample', 7777),
('tample', 8888)])
{'sample': 6666, 'wample': 7777, 'tample': 8888}
```

The `dict()` function generates dictionaries from main sequences and tuple pairs 1 directly.

8.4 Dictionaries with Built-in Functions

The dictionary can have several built-in functions (Table 8.1). In a dictionary the main operations saves a value for the key and retrieves the value for the key [61].

TABLE 8.1

Built-in Dictionary Functions

Built-in Functions	Description
len()	The len() function returns the number of items (key:value pairs) in a dictionary.
all()	The all() function returns a Boolean True value if all the keys in the dictionary are True, otherwise it returns False.
any()	The any() function returns a Boolean True value if any of the keys in the dictionary is True, otherwise it returns False.
sorted()	The sorted() function by default returns a list of items, which are sorted based on dictionary keys.

For example:

```
1. >>>countries = {"USA":1223, "France":1335,
"Brazil":1445, "Italy":1552, "UAE":2445}
2. >>>len(countries)
3. >>>all_dict_func = {0:True, 2:False}
4. >>>all(all_dict_func)
False
5. >>>all_dict_func = {1:True, 2:False}
6. >>>all(all_dict_func)
True
7. >>>any_dict_func = {1:True, 2:False}
8. >>>any(any_dict_func)
True
9. >>>sorted (countries)
["Brazil", "France", , "Italy", "USA", "UAE"]
10. >>>sorted (countries, reverse = True)
["UAE", "USA", , "Italy", "France", "Brazil"]
11. >>>sorted(countries.values())
[1223, 1335, 1445, 1552, 2445]
12. >>>sorted(countries.items())
[('USA', 1223), ('France', 1335), ('Brazil',
1445),('Italy', 1552), ('UAE', 2445)]
```

In the dictionary countries 1 you can find the number of key: value pairs with a function len (2). Every value not null is valid in Python, while null is interpreted as false 2–6. If the Boolean values for all keys are valid, the output in all dictionaries is correct, otherwise there is a false Boolean value. If all of the keys are true, the fact is that there is another False Boolean value in dictionaries 7–8 for any() function. The sorted() function returns the sorted key list in ascending order by default without changing the initial `key: value` pairs 9. The second statement can also be transferred to `reverse = True` 10, which is a sequence of keys ordered in descending order. For dictionary keys that are category strings, their ASCII values are sorted. The `values()` form and the dictionary name 11 can be used to produce an ordered list of values instead of keys. The list of ordered keys, tuple pairs of values, order based on keys and the dictionary name 12 is done using the `items()` process.

8.5 Dictionary Methods

The dictionary helps you to save data without indexing in `key: value` format [62]. Bypassing the `dict` function to `dir`, you can get a list of all `dict` related methods (see Table 8.2).

```
1. >>>dir(dict)
['__class__', '__contains__', '__delattr__', '__
delitem__', '__dir__', '__doc__', '__eq__', '__format__',
'__ge__', '__getattribute__', '__getitem__', '__
gt__', '__hash__', '__init__', '__init_subclass__',
'__iter__', '__le__', '__len__', '__lt__', '__ne__',
'__new__', '__reduce__', '__reduce_ex__', '__repr__',
'__setattr__', '__setitem__', '__sizeof__', '__str__',
'__subclasshook__', 'clear', 'copy', 'fromkeys', 'get',
'items', 'keys', 'pop','popitem', 'setdefault', 'update',
'values']
```

TABLE 8.2

Different Types of Dictionary Methods

Dictionary Method	Syntax	Description
clear()	dictionary_name. clear()	The clear() method removes all the key:value pairs from the dictionary.
fromkeys()	dictionary_name. fromkeys(seq [, value])	The fromkeys() method creates a new dictionary from the given sequence of elements with a value provided by the user.
get()	dictionary_name. get(key[, default])	The get() method returns the value associated with the specified key in the dictionary. If the key is not present then it returns the default value. If the default is not given, it defaults to None, so that this method never raises a KeyError.

(Continued)

TABLE 8.2 (CONTINUED)

Different Types of Dictionary Methods

Dictionary Method	Syntax	Description
items()	dictionary_name. items()	The items() method returns a new view of the dictionary's key and value pairs as tuples.
keys()	dictionary_name. keys()	The keys() method returns a new view consisting of all the keys in the dictionary.
pop()	dictionary_name. pop(key[, default])	The pop() method removes the key from the dictionary and returns its value. If the key is not present, then it returns the default value. If default is not given and the key is not in the dictionary, then it results in KeyError.
popitem()	dictionary_name. popitem()	The popitem() method removes and returns an arbitrary (key, value) tuple pair from the dictionary. If the dictionary is empty, then calling popitem() results in KeyError.
setdefault()	dictionary_name. setdefault (key[, default])	The setdefault() method returns a value for the key present in the dictionary. If the key is not present, then insert the key into the dictionary with a default value and return the default value. If key is present, default defaults to None, so that this method never raises a KeyError.
update()	dictionary_name. update([other])	The update() method updates the dictionary with the key:value pairs from other dictionary objects and returns None.
values()	dictionary_name. values()	The values() method returns a new view consisting of all the values in the dictionary.

For example:

```
1. >>>box_office_billion = {"Bahubali":2009,
"3idiot":1997, "Holiday":2015, "harrypotter":2011,
"avengers":2012}
2. >>>box_office_billion_fromkeys = box_office_billion.
fromkeys(box_office_billion)
3. >>>box_office_billion_fromkeys
{'Bahubali': None, '3idiot': None, 'Holiday': None,
'harrypotter': None, 'avengers': None}
4. >>>box_office_billion_fromkeys = box_office_billion.
fromkeys(box_office_
billion, "billion_dollar")
5. >>>box_office_billion_fromkeys
{'Bahubali': 'billion_dollar', '3idiot': 'billion_
dollar', 'Holiday': 'billion_dollar', 'harrypotter':
'billion_dollar', 'avengers': 'billion_dollar'}
```

```
6. >>>print(box_office_billion.get("frozen"))
None
7. >>>box_office_billion.get("frozen",2013)
2013
8. >>>box_office_billion.keys()
dict_keys(['Bahubali', '3idiot', 'Holiday',
'harrypotter', 'avengers'])
9. >>>box_office_billion.values()
dict_values([2009, 1997, 2015, 2011, 2012])
10. >>>box_office_billion.items()
dict_items([('Bahubali', 2009), ('3idiot', 1997),
('Holiday', 2015), ('harrypotter', 2011),
('avengers', 2012)])
11. >>>box_office_billion.update({"frozen":2013})
12. >>>box_office_billion
{'Bahubali': 2009, '3idiot': 1997, 'Holiday': 2015,
'harrypotter': 2011, 'avengers': 2012, ' frozen': 2013}
13. >>>box_office_billion.setdefault("minions")
14. >>>box_office_billion
{'Bahubali': 2009, '3idiot': 1997, 'Holiday': 2015,
'harrypotter': 2011, 'avengers': 2012,
' frozen': 2013, 'minions': None}
15. >>>box_office_billion.setdefault("ironman", 2013)
16. >>>box_office_billion
{'Bahubali': 2009, '3idiot': 1997, 'Holiday': 2015,
'harrypotter': 2011, 'avengers': 2012,
'minions': None, 'ironman': 2013}
17. >>>box_office_billion.pop("Bahubali")
2009
18. >>>box_office_billion.popitem()
('ironman', 2013)
19. >>>box_office_billion.clear()
{}
```

Diverse dictionary processing is done by the dictionary methods 1–19. Dict keys, dict values, and dict object data types are read-only viewpoints that cannot be modified directly from separate dictionary methods. If you choose to translate dict keys, dict values and dict points data types retrieved from keys(), values(), and items() to a true list, the returns list would be moved to list() function returns values. Whenever the vocabulary evolves, the opinions of certain data types are represented. For examples:

```
1. >>>list(box_office_billion_keys())
['Babubali', 3idiot', 'Holiday', harrypotter',
   'avengers']
```

```
2. >>list(box_office_billion.values())
[2009, 1997,2015,2011,2012]
3. >>>list(box_office_billion.item())
[('Bahubali', 2009), ('3idiot', 1997), ('Holiday', 2015),
    ('harrypotter', 2011), ('avengers', 2012)]
```

The dict-key, dict-word and dict items using list()function 1– 3, the keys (), values() and objects() returned are transformed into a true list.

8.5.1 Population of Primary Dictionaries: Value Pairs

One easy way to construct dictionaries is to start using an empty{} dictionary in order to add a key value using update(). If there is no key, then immediately the key: value pair is generated and attached to the dictionary.
   For example:

```
1. >>>countries = {}
2. >>>countries.update({"Asia":"Sri lanka"})
3. >>>countries.update({"Europe":"France"})
4. >>>countries.update({"Africa":"Ghana"})
5. >>>countries
{'Asia': 'Sri lanka', 'Europe': 'France', 'Africa':
'Ghana'}
```

Let countries empty of the dictionary 1 and begin to add the key to the dictionary by using the 2–4 update() feature and display the key to the dictionary countries: dictionary countries value pairs 5. Add the value pair of curly braces as the key function update().

8.5.2 Dictionary Traversing

A loop should be used only to iterate over keys, values or entities. When using a loop to iterate through a dictionary, you iterate through the keys by default. For repeating the value() method and the value pairs: the latter specifically define the item() method from the dictionary if the values are to be iterated. You use dictionary keys, dictation values and dictionaries to iterate using key pairs or value pairs returned by the dictionary process [63].

```
currency = {"India": "Rupee", "USA": "Dolla r", "Russia":
"Ruble", "Japan": "Yen", "Germany": "Euro"}
```

```
def main():
  print("List of Countries")
  for key in currency.keys():
    print(key)
    print("List of Currencies in different Countries")
  for value in currency.values():
    print(value)
  for key, value in currency.items():
    print(f"'{key}' has a currency of type '{value}'")
  if__name__ == "__main__":
    main()
```

Output

```
List of Countries
India
List of Currencies in different Countries
USA
List of Currencies in different Countries
Russia
List of Currencies in different Countries
Japan
List of Currencies in different Countries
Germany
List of Currencies in different Countries
Rupee
Dollar
Ruble
Yen
Euro
'India' has a currency of type 'Rupee'
'USA' has a currency of type 'Dollar'
'Russia' has a currency of type 'Ruble'
"Japan' has a currency of type 'Yen'
'Germany' has a currency of type 'Euro'
```

The keys() 4–5, values() 7–8, and objects() 9–10 are used to allow a loop to iterate over the keys, values or objects. A key is iterated by default for a loop. Then the declaration for the currency base. Key(): output results in the same currency output as the key. The key and its accompanying value can be obtained simultaneously with the object() process while looping via dictionaries. The values returned by the objects() method in the dict items form are multiples, where the first element in the multiple is the reference, the second is the value.

8.6 The Del Statement

Use the del button, accompanied by the dictionary name and the key you want to extract, to uninstall the key: value pair [62].

```
del dict_name[key]
```

```
1. >>>animals = {"r":"raccoon"; "c"':"cougar".
"m":"moose"}
2. >>>animals
{'r': 'raccoon', 'c': 'cougar', 'm': 'moose'}
3. >>>del animals["c"]
4. >>>animals
{'r': 'raccoon', 'm': 'moose'}
The key: value pair of "c" can be deleted in the animals
1-2 dictionary, as seen in 3: "cougar."
```

8.7 Conclusion

The Python dictionary is a list that looks like a map to hold key value pairs. The articles are accessible via a main index in the dictionary. Dictionary objects are quickly modified, inserted and deleted. There are different ways to iterate through dictionary keys, values or objects for loops. The main properties of the dictionary have been discussed in this chapter, explaining how to access the dictionary items in the list. One of the most popular Python concepts are lists and dictionaries. The list items are accessible by order-based numerical index, and are key-based in dictionary items. As a result, lists and dictionaries tend to fit various types of conditions.

9

Tuples and Sets

9.1 Creating Tuples

A tuple is a limited orderly arrangement of potentially different types of values, used for bundling similar values without requiring a particular form to store them [64]. Tuples are unchanging. You cannot change the values once a tuple is formed. A tuple is set by using parentheses () with a comma-separated array of values. In a tuple, any value is referred to as an object. The tuple generation syntax is:

tuple-name = {value-1, value-2, value- 3, ..., value-n}

For example:

```
1. >>>program = ("C", "Dennis Ritchie", "Language", 1972)
2. >>> program
('C', ' Dennis Ritchie,' ' Language ', 1972)
3. >>>type(internet)
<class 'tuple'>
4. >>>Cars = "ferrari", "tata motors", "mercedes", "MG",
"renault"
5. >>>Cars
('ferrari', 'tata motors', 'mercedes', 'MG', 'renault')
6. >>>t ype(cars)
<class 'tuple'>
```

In 1, a list of numbers and string forms are added to the tuple domain. The tuple contents are shown with the Program 2 tuple tag. The contents and order of the things in the folding are the same as when the folding is made. The tuple class is called "tuple 3" in Python. Tuples are also stored in parentheses so that they are correctly displayed. Tuples may be made without parentheses. It is the comma that really makes the tuples, rather than parentheses 4-6. You can create an empty tuple without any numbers. The syntax is:

$$tuple\_name = ()$$

For example:

```
1. >>> empty_tuple = ()
2. >>> empty_tuple()
3. >>> type(empty_tuple)
<class 'tuple=>
As seen in 1 an open tuple is generated and an empty
tuple is of form 3 tuples.
For examples,
1. >>>air_force = ("f16", "f10a", "f88a")
2. >>>fighter_jets = (1990, 2009, 2017, air_force)
3. >>>fighter_jets
(1990, 2009, 2017, ('f16', 'f10a', 'f88a'))
4. >>>type(fighter_jets)
<class 'tuple'>
```

In this case, airforce is tuple form 1. The tuple fighter jets 4 is made up of a set of integer forms and a tuple 2. The building of tuples containing 0 or 1, which have some additional characteristics to handle, is a special challenge. For example:

```
1. >>>empty = ()
2. >>>singleton = 'fine'
3. >>>singleton
('fine')
```

A null pair of parentheses are inserted into null tuples 1. A tuple of one object has a meaning followed by a comma 2. A single value in parentheses is not appropriate.

9.2 Basic Tuple Operations

As with lists, the + operator is used to merge tuples and the * operator is used in repeater tuple series [65]. For example:

```
1. >>>tuples_1 = (2, 0, 1, 5)
2. >>>tuples_2 = (2, 0, 2, 0)
3. >>>tuples_1 +tuples_2
(2, 0, 1, 5, 2, 0, 2, 0)
4. >>>tuples_1 * 3
(2, 0, 1, 5, 2, 0, 1, 5, 2, 0, 1, 5)
5. >>>tuples_1 == tuples_2
False
```

Two tuples, tuple 1 and tuple 2, are formed from 1 to 2. To create a new tuple, you add tuple 1 and tuple 2. All the objects in the current tuple are added to tuple 3. The multiplication function * can be used on tuples. The articles in the tuples are repeated as many times as you decide and three times in the fourth tuple 1 list. In 5, the tuple objects are compared with the operator = =, and the answer is Boolean False as the tuples are distinct.

You can verify the existence of an object in a tuple. It s a True or False Boolean. For example:

```
1. >>>tuple_items = (1, 9, 9, 9)
2.>>>1 in tuple_items
True
3. >>>25 in tuple_items
False
```

If an object is present in tuple 2, the operator returns Boolean True, or False Boolean 3.

<, < =, >, > =, = = and the equate operators ! = are used for tuple analysis. For example:

```
1. >>>tuple_1 = (9, 8, 7)
2. >>>tuple_2 = (9, 1, 1)
3. >>>tuple_1 >tuple_2
True
4. >>>tuple_1!=tuple_2
True
```

Place by location is contrasted with tuples. The first point of the first tuple is compared to the second tuple. If it is not the same, it is compared with the second element of the second tuple of the first tuple to see whether it is not equivalent, it is the comparable attribute, or is the third element considered 1-4.

9.2.1 The tuple () Function

The combined tuple() function is used for tuples [66]. The syntax is tuple type:

$$([sequence])$$

where a number, string or tuple is the series. If you do not define the optional sequence, an empty tuple will be generated.

```
1.  >>>fresh = "ziQueen"
2.  >>>string_to_tuple = tuple(fresh)
3.  >>>string_to_tuple
( 'z', 'i', 'Q', 'u', 'e', 'e', 'n')
4.  >>>massenger = ["g", "o", "d", "o", "f", "o", "n",
"e"]
5.  >>>list_to_tuple = tuple(messenger)
6.  >>>list_to_tuple
( 'g', 'o', 'd', 'o', 'f ', 'o', 'n', 'e')
7.  >>>string_to_tuple +"scandinavia"
            Traceback (most recent call last):
            File "<stdin>", line 1, in <module>
            TypeError: can only concatenate tuple (not
"str") to tuple
8.  >>>string_to_tuple +tuple("scandinavia")
( 'z', 'i', 'Q', 'u', 'e', 'e', 'n', 's', 'c', 'a', 'n',
'd', 'i', 'n', 'a', 'v', 'i', 'a')
9.  >>>list_to_tuple +[ "b", "r", "e", "e", " k"]
Traceback (most recent call last):
File "<stdin>", line 1, in <module>
TypeError: can only concatenate tuple (not "list") to
tuple
10. >>>list_to_tuple +tuple([ "b", "r", "e", "e", " k"])
( 'g', 'o', 'd', 'o', 'f ', 'o', 'n', 'e', 'b', 'r', 'e',
'e', 'k')
11. >>>letters = ("p", "q", "r")
12. >>>numbers = (4, 5, 6)
13. >>>nested_tuples = (letters, numbers)
14. >>>nested_tuples
(('p', 'q', 'r'), (4, 5, 6))
15. >>>tuple("beautiful")
( 'b', 'e', 'a', 'u', 't', 'i', 'f', 'u', 'l')
```

The string variable 1 has been transformed to a tuple with the feature `tuple ()` 2. It is not only possible to transform strings, but even list 4 variables to tuples

5. When you try to connect either a string 7 or a tuple list 9, this will lead to an error. Before you concatenate with tuple forms 8 and 10, you can convert strings and lists to tuples using the `tuple()` function. Python is approved to assemble tuples. A tuple process output is often stored in parentheses such that nesting tuples are properly represented 11–14. When a string is translated to a tuple 15, a comma divides a single character in a series.

9.3 Indexing and Slicing in Tuples

An index may be used for each object in a tuple [67]. The word in the brackets is referred to as the table. Square brackets [] are used for tuples, with the first object being indexed to 0, the second to 1, and so on. This index shows the number that is accessed in the square brackets. The syntax to enter a tuple object is:

$$tuple\_name[index]$$

where the index often needs to be an integer value that shows which object to pick. The index description is shown below for the tuple's holy places:

Temple	Vashino Devi	Jawala Maa	Tarnote Maa	Triupati Balaji	Saibaba
	0	1	2	3	4

For example,

```
1. >>>temple = ("Vashino Devi", "Jawala Maa", "Tarnote
Maa", "Triupati Balaji", "Saibaba")
2. >>>temple
("Vashino Devi", "Jawala Maa", "Tarnote Maa", "Triupati
Balaji", "Saibaba")
3. >>>temple[0]
' Vashino Devi '
4. >>>temple[1]
' Jawala Maa '
5. >>>temple[2]
' Tarnote Maa '
6. >>>temple[3]
' Triupati Balaji '
7. >>>temple[4]
' Saibaba '
8. >>>temple[5]
Traceback (most recent call last):
File "<stdin>", line 1, in <module>
IndexError: tuple index out of range
```

In the tuple, the first object is shown directly after the tuple with a value of zero 1 using the square brackets. This tuple is made up of 0–4 index numbers. The resulting error is "IndexError: tuple index out of control" where a value indexed exceeds the number of tuple items 8 [68].

Alongside positive index values, you can also enter tuple items with a negative index number at −1 starting at the end of the tuple. If you have a lot of things in the tuple and want to find an entry in the tuple, negative indexing is beneficial. The negative index description occurs below for the same tuple's holy places:

Temple	Vashino Devi	Jawala Maa	Tarnote Maa	Triupati Balaji	Saibaba
	−5	−4	−3	−2	−1

For example:

```
1. >>>holy_places[-2]
' Triupati Balaji '
```

You can use its negative index count to print out the item "Bethlehem", as in 1.

Tuples are permitted to slice in Python in which a part of the tuple can be extracted with the column (:) operator by setting an index range which results in a tuple form.

The tuple trimming syntax is:

```
tuple-name[start:stop[:step]]
```

where the integer (positive or negative value) is both the start and end. The tuple slice returns a tuple component to an index key that includes a start index value but does not have the value of the stop index. The move determines the slice raise value and is optional.

The positive and negative index colors are shown below for the tuple colors:

Colors	"B"	"G"	"R"	"Y"	"O"	"P"	"M"
	0	1	2	3	4	5	6
	−7	−6	−5	−4	−3	−2	−1

For example:

```
1. >>>colors = ("B", "G", "R", "Y", "O", "P", "M")
2. >>>colors
( 'B', 'G', ' R', 'Y', 'O', 'P', 'M')
3. >>>colors[1:4]
('G', ' R', 'Y')
4. >>>colors[:5]
('B', 'G', ' R', 'Y', 'O')
5. >>>colors[3:]
( 'Y', 'O', 'P', 'M')
6. >>>colors[:]
( 'B', 'G', ' R', 'Y', 'O', 'P', 'M')
7. >>>colors[::]
 ( 'B', 'G', ' R', 'Y', 'O', 'P', 'M')
8. >>>colors[1:5:2]
( 'G', 'Y')
9. >>>colors[::2]
( 'B', ' R', 'O', 'M')
10. >>>colors[::-1]
('M', 'P', 'O', 'Y', 'R', 'G', 'B')
11. >>>colors[-5:-2]
('R', 'Y', 'O')
```

The tuple colors have seven-string elements. All color articles are compounded by 1 to 4, but only 4 and 3 are sliced. All color articles are numerical. If from the beginning of the index you want to reach multiple objects, no value of zero should be specified. The index value may be skipped and only the stop index value 4 is defined. Even if you want to allow tuple access to the items up to the end of the tuple, then the stop value is not necessary, and only the start index value 5 should be listed. If you ignore the start and finish index value 6 and just mention the colon consumer within the parentheses, then all tuple objects will be seen. Instead of using a singular colon, the entire output of tuple 7 is seen too. The second column statement indicates to Python that you want to select an increase in slicing. By default, Python sets this number to 1; however, this number may be defined after the second column to 8. The use of a double-column as shown in 7 implies no beginning index value and no holdback index value and bounces the objects in two steps. Any second tuple object is derived from a zero index value. In the reverse order, the double column and an index value of 1 to 10 are shown for all things in a tuple. For start and end index values, 11 negative index values may also be used.

9.4 Built-in Functions of Tuples

Many built-in functions (Table 9.1) can be used to transmit a tuple as an argument [69].

TABLE 9.1

Built-in Functions in Tuples

Built-in Functions	Description
len()	The len() function returns the numbers of items in a tuple.
sum()	The sum() function returns the sum of the numbers in the tuple.
sorted()	The sorted() function returns a sorted copy of the tuple as a list while leaving the original tuple untouched.

For example:

```
1. >>>years = (1988, 1986, 1983, 1999)
2. >>>len(years)
4
3. >>>sum(years)
7956
4. >>>sorted_years = sorted(years)
5. >>>sorted_years
[1983, 1986, 1988, 1999]
```

In double years the function len() 2 can be used to locate the number of products. The sum() function gives the added value of any number object in tuple 3. Without altering a tuple of the original object allocated to the current list vector 4, sorted() returns a list of the classified items. In the case of double string products, ASCII values are graded.

9.5 Comparison Between Tuples and Lists

Tuples can look like lists but are also used for various contexts and purposes [70]. Tuples are unmodified and usually have a heterogeneous string of elements that are unpackaged or indexed. Lists may be modified and indexed for their items. In a tuple, no items can be added, omitted or replaced. For example:

```
1. >>>coral_reel = ("great", "ningaloo", "amazon",
"pickles")
2. >>>coral_reel[0] = "pickles"
Traceback (most recent call last):
File "<stdin>", line 1, in <module>
TypeError: 'tuple' object does not support item
assignment
3. >>>coral_reel_list = list(coral_reel)
4. >>>coral_reel_list
['great', 'ningaloo', 'amazon', 'pickles']
```

When you try replacing the "great barrier" in tuple coral reel by another object, such as "pickles", TypeError fails, because tuples can't be changed 1–2. Bypassing the tuple name to the list() function 3–4, allows you to transform the tuple into a number.

You can modify an object if a tuple is mutable. If a collection is an object in a tuple, then any modifications in the tuple category will be mirrored in the total products. For example:

```
1. >>>german_cars = ["Porsche car", "audi car", "bmwcar"]
2. >>>Indian_cars = ("Manuti", "Tata", "Mahindra",
german_cars)
3. >>>Indian_cars
('Maruti', 'Tata', 'mahindra', ['porsche', 'audi',
'bmw'])
4. >>>greman_cars[3].append("mercedes")
5. >>>german_cars
['porsche', 'audi', 'bmw', 'mercedes']
6. >>>indian_cars
('Maruti', 'Tata', 'mahindra', ['porsche', 'audi', 'bmw',
'mercedes'])
```

When the underlying list varies, the tuple "containing" a number tends to be changed. The main idea is that tuples can't tell whether the objects within them are mutable. The only thing that can modify an object is a process that changes the results. There is no way that this can be observed in general. The tuple has not necessarily altered. It cannot be adjusted and there are no mutating processes. The tuple is not informed of updates until the list changes. The list does not know whether a vector, a tuple or a separate set is applied to 1–6. The tuple is not mutable, so it has no way to alter its contents. Also, the string is unchanging since strings have no form of mutation.

9.6 Comparsion Between Tuples and Dictionaries

Tuples may be used for constructing dictionaries as key: value pairs [71]. For example:

```
1. >>>fish_weight_kg = (("white_whale", 530), ("small
whale", 1581), ("Greenland_whale", 1410))
2. >>>fish_weight_kg_dict = dict(fish_weight_kg)
3. >>>fish_weight_kg_dict
{'white_whale ': 530, 'small whale': 1581, 'greenland_
whale': 1410}
```

By transferring the tuple name to a dictation function, the tuples can be translated to dictionaries. This is accomplished by nesting a tuple in a tuple, with each nested tuple object 1-2 containing two items. When the tuple is translated to a dictionary 3, the first element becomes the key, and the second component its meaning.

The item() method returns the tuple list in a dictionary, where every tuple matches a key: a dictionary value pair. For example:

```
1. >>>company_build_year = {"Google":1996, "Apple":1976,
"Sony":1946, "ebay":1995, "IBM":1911}
2. >>>founding_build_year.items()
dict_items([('Google', 1996), ('Apple', 1976), ('Sony',
1946), ('ebay', 1995), ('IBM', 1911)])
3. >>>for company, year in founding_build_year.items():
... print(f"{company} was found in the year {year}")
```

Output

```
Google was found in the year 1996
Apple was found in the year 1976
Sony was found in the year 1946
ebay was found in the year 1995
IBM was found in the year 1911
```

There are two parameters to the loop in 3: business and year when the items() process restores a fresh perspective on the key and value pairs on each key: the dictionary value pairs.

9.7 Tuple Methods

By moving the tuple function on to `dir()`, you can get a list of all methods relevant to the tuple (Table 9.2):

```
1. >>>dir(tuple)
['__add__', '__class__', '__contains__', '__delattr__',
'__dir__', '__doc__', '__eq__', '__format__', '__ge__',
'__getattribute__', '__getitem__', '__getnewargs__',
'__gt__', '__hash__', '__init__', '__init_subclass__',
'__iter__', '__le__', '__len__', '__lt__', '__mul__',
'__ne__', '__new__', '__reduce__', '__reduce_ex__', '__
repr__', '__rmul__', '__setattr__', '__sizeof__', '__
str__', '__subclasshook__', 'count', 'index']
```

For example:

```
1. >>>Broadcast_channels = ("sab", "discovery", "animal_
planet", "starsports", "sony")
2. >>> Broadcast_channels.count("sab")
2
3. >>> Broadcast_channels.index("starsports ")
3
```

TABLE 9.2

Different Methods of Tuples

Tuple Methods	Syntax	Description
count()	tuple_name.count(item)	The count() method counts the number of times the item has occurred in the tuple and returns it.
index()	tuple_name.index(item)	The index() method searches for the given item from the start of the tuple and returns its index. If the value appears more than once, you will get the index of the first one. If the item is not present in the tuple, then ValueError is thrown by this method.

9.7.1 Tuple Packing and Unpacking

The declaration s = 56789, 98765, 'hi! 'It's a tuple packaging proof [72].

```
1. >>>s = 56789, 98765, 'hi!'
2. >>>s
(56789, 98765, 'hi!')
The 56789 value, 98765 values, hi! 'They're bundled into
a 1-2 tuple together.
It is also possible to remove the tuple wrapping. For
examples,
1. >>>p, q, r = s
2. >>>p
56789
3. >>>q
98765
4. >>>r
'hello!'
```

The procedure is referred to as a tuple unpack which operates on every list on the right. Tuple unpackage includes that on the left hand of the same sign; as in the tuple 1 to 4 there are almost as many variables. Remember that some functions are just a mixture of tuple packaging and unpackaging.

9.7.2 Tuples Traversing

You can iterate in loops for any object [73]:

```
ocean_animals = ("electric_eel","jelly_fish","shrimp",
"turtles","blue_whale")
def main():
    for each_animal in ocean_animals:
        print(f"{each_animal} is an ocean animal")
if __name__ == "__main__":
    main()
```

Output

```
electric_eel is an ocean animal
jelly_fish is an ocean animal
shrimp is an ocean animal
turtle is an ocean animal
blue_whale is an ocean animal
```

Here, tuple category 1 is ocean animals. A three-step loop is being used to iterate a tuple object.

9.7.3 Tuples with Items

You can use the + = operator and also transform list items into tuples of elements [70].

```
tuple_items = ()
total_items = int(input("Enter the total number of items: "))
for i in range(total_items):
    user_input = int(input("Enter a number: "))
    tuple_items += (user_input,)
    print(f"Items added to tuple are {tuple_items}")
    list_items = []
    total_items = int(input("Enter the total number of items:
        "))
for i in range(total_items):
    item = input("Enter an item to add: ")
    list_items.append(item)
    items_of_tuple = tuple(list_items)
    print(f"Tuple items are {items_of_tuple}")
```

Output

```
Enter the total number of items: 4
Enter a number: 4
Items added to tuple are (4,)
Enter the total number of items: 3
Enter a number: 3
Items added to tuple are (4, 3)
Enter the total number of items: 2
Enter a number: 2
Items added to tuple are (4, 3, 2)
Enter the total number of items: 1
Enter a number: 1
Items added to tuple are (4, 3, 2, 1)
Enter the total number of items: 4
Enter an item to add: 6
Tuple items are ('6',)
Enter an item to add: 4
Tuple items are ('6', '4')
Enter an item to add: 9
Tuple items are ('6', '4', '9')
Enter an item to add: 8
Tuple items are ('6', '4', '9', '8')
```

Two approaches are used to insert objects into the tuple: using a fixed + = operator 1–6 and by transforming list items to tuple items 7–13. Tuple items are tuple-type in the language. In each of the methods, the number of things to be applied to the tuple should be defined in advance 2, 8. On this basis, we use the range (function) 3–4 and 9–10 as much as possible via the for loop. In the first step, the consumer enters tuples continuously with the + = operator. Tuples are immutable and cannot be modified. Any original + (new element) is substituted throughout each iteration, generating a new tuple 5. List 7 is generated in the second process. The user inserted value 10 is added to the list variable for each iteration 9. This array is translated to a tuple with the feature 12 of tuple.

9.8 Use of the Zip() Function

The `zip()` function produces an element series from each iterable (can be zero or more) [68]. The `zip()` function syntax is:

$$zip(*iterables)$$

A number, string or dictionary may be iterable. It yields a number of tuples in which each iterable element is used. The product aggregation is stopped if the shortest iterable input is consumed. For example, the `zip` function returns tuples twice when two iterables are moved, one containing two objects, one containing five more. It returns an iterator with a single iterable declaration. It returns a null iterator without any arguments. For example:

```
1. >>>w = [7, 8, 9]
2. >>>yz= [4, 5, 6]
3. >>>zipped = zip(w, yz)
4. >>>list(zipped)
[(7,4), (8, 5), (9, 6)]
```

Here `zip()` is used to zip two array 1–4 iterables. The entries may be combined with the `zip()` function to loop over two or more sequences simultaneously. For example:

```
1. >>>questions = ('name ', 'quest', 'favourite colour')
2. >>>answers = ('Binny', 'the holy fest', 'green')
3. >>>for q, a in zip(questions, answers):
... print(f'What is your {q}? It is {a}.')
```

Output

```
What is your name? It is Binny.
What is your quest? It is the holy fest.
What is your favourite color? It is green.
```

Because zip () returns the tuple value, you can print tuple 1–3 with a loop with many iterating variables.

9.9 Python Sets

Python also incorporates sets of a data form. A package has no replication of an unordered array. Checking memberships and deleting redundant entries are the key uses of packages. Systems also help statistics, including union, conjunction, variance and symmetrical disparities [71].

Curly braces {} or set() can be used to construct sets in curly brackets with a comma-separated list of {} elements. Note: you must use set() to create a blank list rather than {} since the latter creates a static dictionary.

```
1.  >>>baskets = {'apples', 'oranges', 'apples', 'pears',
'oranges', 'bananas'}
2.  >>>print(baskets)
{'bananas', 'apples', 'oranges', 'pears'}
3.  >>>'oranges' in baskets
True
4.  >>>'crabgrass' in baskets
False
5.  >>>c = set('abracadabra')
6.  >>>d = set('alacazam')
7.  >>>c
{'d', 'a', ' b', 'r', 'c'}
8.  >>>d
{'m', 'l', 'c', 'z', 'a'}
9.  >>>c - d
{' b', 'r', 'd'}
10. >>>c | d
{'l', 'm', 'z', 'd', 'a', ' b', 'r', 'c'}
11. >>>c & d
{'a', 'c'}
12. >>>c ^ d
{'l', 'd', 'm', ' b', 'r', 'z'}
13. >>>len(baskets)
4
14. >>>sorted(baskets)
['apples', 'bananas', 'oranges', 'pears']
```

A selection of specific objects is a series. Duplicate products from the set basket are excluded 1. While the items "oranges" and "apples" have been presented two times, there is only one item from "oranges", "apples" and "pears" that will be included in the package. Sets a and b reveal identical letters 5–8. Set a but not set b letters are written 9. Set c, set d, or all of these letters are written in 10. Both letters have a bare print and collection. Letters in set d are written 12, but not both. The lens function 13 includes a cumulative number of objects in the fixed bin. Sorted() returns a new sorted list of set 14 items.

9.10 Set Methods

By bypassing the set function to dir(), you can obtain a list of all procedures connected to the set (Table 9.3).

```
1. >>>dir(set)
['__and__', '__class__', '__contains__', '__delattr__',
'__dir__', '__doc__', '__eq__', '__format__', '__ge__',
'__getattribute__', '__gt__', '__hash__', '__iand__',
'__init__', '__init_subclass__', '__ior__', '__isub__',
'__iter__', '__ixor__', '__le__', '__len__', '__lt__',
'__ne__', '__new__', '__or__', '__rand__', '__reduce__',
'__reduce_ex__', '__repr__', '__ror__', '__rsub__',
'__rxor__', '__setattr__', '__sizeof__', '__str__', '__
sub__', '__subclasshook__', '__xor__', 'add', 'clear',
'copy', 'difference', 'difference_update', 'discard',
'intersection', 'intersection_update', 'isdisjoint',
'issubset', 'issuperset', 'pop', 'remove', 'symmetric_
difference', 'symmetric_difference_update', 'union',
'update']
```

TABLE 9.3

Different Set Methods

Set Methods	Syntax	Description
add()	set_name.add(*item*)	The add() method adds an item to the set set_name.
clear()	set_name.clear()	With the clear() process, the set set_name eliminates all objects.
difference()	set_name. difference(*others*)	The difference() method generates a string set of elements that are not set in the other sets in set_name.

(Continued)

TABLE 9.3 (CONTINUED)

Set Methods	Syntax	Description
discard()	set_name. discard(item)	If this is present, the discard() method excludes an object from set_name.
intersection()	set_name. intersection(*others*)	The function intersection() returns a new set of objects common to the sets.
isdisjoint()	set_name. isdisjoint(*other*)	The isdisjoint() returns True if there is no common object in the set set_name. Sets would be disjointed only if the crossroads are zero.
issubset()	set_name. issubset(*other*)	If any object in the set set_name is set in another set, the emuset() method returns True.
issuperset()	set_name. issuperset(*other*)	The issuperset() function returns True if the set name contains all the elements of another set.
pop()	set_name.pop()	The pop() method deletes an undefined object from set_name and returns it. If the set is empty, it raises KeyError.
remove()	set_name. remove(*item*)	Removes the object from set_name using the remove() process. If the object isn't in the set, it elevates KeyError.
symmetric_ difference()	set_name. symmetric_ difference(other)	A new set of elements either in the set or in another set, but not both, comes with the form symmetric difference().
union()	set_name. union(*others*)	A new set of elements of set_name and all other sets is returned by method union().
update()	set_name. update(*others*)	Update the set_name by inserting objects from all collections.

```
1. >>>Indian_flowers = {"sunflowers", "roses",
"lavender", "tulips", "goldcrest"}
2. >>>american_flowers = {"roses", "tulips", "lilies",
"daisies"}
3. >>>american_flowers.add("orchids")
4. >>>american_flowers.difference(Indian_flowers)
{'lilies', 'orchids', 'daisies'}
5. >>>american_flowers.intersection(Indian_flowers)
{'roses', 'tulips'}
6. >>>american_flowers.isdisjoint(Indian_flowers)
False
7. >>>american_flowers.issuperset(Indian_flowers)
False
8. >>>american_flowers.issubset(Indian_flowers)
False
```

```
9. >>>american_flowers.
symmetric_difference(Indian_flowers)
{'lilies', 'orchids', 'daisies', 'goldcrest',
'sunflowers', 'lavender'}
10. >>>american_flowers.union(Indian_flowers)
{'lilies', 'tulips', 'orchids', 'sunflowers', 'lavender',
'roses', 'goldcrest', 'daisies'}
11. >>>american_flowers.update(Indian_flowers)
12. >>>american_flowers
{'lilies', 'tulips', 'orchids', 'sunflowers', 'lavender',
'roses', 'goldcrest', 'daisies'}
13. >>>american_flowers.discard("roses")
14. >>>american_flowers
{'lilies', 'tulips', 'orchids', 'daisies'}
15. >>>Indian_flowers.pop()
'tulips'
16. >>>american_flowers.clear()
17. >>>american_flowers
set()
```

Different sets are managed by specified methods 1–17.

9.10.1 Traversing of Sets

```
warships = {"u.s.s._arizona", "hms_beagle", "ins_airavat",
"ins_hetz"}
def main():
    for each_ship in warships:
        print(f"{each_ship} is a Warship")
if__name__ == "__main__":
    main()
```

Output

```
u.s.s._arizona is a Warship
hms_beagle is a Warship
ins_airavat is a Warship
ins_hetz is a Warship
```

Warships here are of a specific form 1. A loop for 3 is used for each object in the collection for iteration.

9.11 The Frozen Set

A frozen set is essentially the same as a package, except it is unchanging. If you create a frozen set, you cannot modify the objects [67]. They can be found in other collections as representatives and as dictionary keys, as they are immutable. Frozen sets have the same features as ordinary sets, with no operation (update, delete, rise, etc.) that alters the contents.

```
1. >>>dir(frozenset)
['__and__', '__class__', '__contains__', '__delattr__',
'__dir__', '__doc__', '__eq__', '__format__', '__ge__',
'__getattribute__', '__gt__', '__hash__', '__init__',
'__init_subclass__', '__iter__', '__le__', '__len__',
'__lt__', '__ne__', '__new__', '__or__', '__rand__',
'__reduce__', '__reduce_ex__', '__repr__', '__ror__',
'__rsub__', '__rxor__', '__setattr__', '__sizeof__', '__
str__', '__sub__', '__subclasshook__', '__xor__', 'copy',
'difference', 'intersection', 'isdisjoint', 'issubset',
'issuperset', 'symmetric_difference', 'union']
```

For example:

```
1. >>>fs = frozenset(["g", "o", "o", "d"])
2. >>>fs
    frozen set ({'d', 'o', g'})
3. >>>animals = set([fs, "cattle", "horse"])
4. >>animals
{'cattle', frozenset({'d', 'o', 'g'}), 'horse'}
5. >>>official_languages_world = {"english":69,
"french":39, "spanish":41}
6. >>>frozenset(official_languages_world)
frozenset({'spanish', 'french', 'english'})
7. >>>frs = frozenset(["german"])
8. >>>official_languages_world = {"english":69,
"french":39, "spanish":41, frs:7}
9. >>>official_languages_world
{'english': 69, 'french': 39, 'spanish':41,
frozenset({'german'}): 7}
```

If a dictionary is sent to frozenset() function 6, the key is returned in a dictionary. Frozenset is a guide in the dictionary 8.

9.12 Conclusion

This chapter has discussed the tuple, which is an immutable data structure comprising items that are ordered and heterogeneous. Tuples are formed using commas and not parentheses. Indexing and slicing of items are supported in tuples. They support built-in functions such as `len()`, `min()` and `max()`. The set stores a collection of unique values which are not placed in any particular order. You can add an item to the set using the `add()` method and remove an item using the `remove()` method. The for loop is used to traverse the items in a set. The `issubset()` or `issuperset()` method is used to test whether a set is a superset or a subset of another set. Sets also provide functions such as `union()`, `intersection()`, `difference()` and `symmetric difference()`.

10

Strings and Special Methods

10.1 Introduction

In almost all types of programs, the use of strings abounds. A string comprises a series of letters, numbers, punching signs and spaces. You may use one quote, a double quote or three quotes to represent lines [74].

10.2 Creating and Storing Strings

Strings are another simple Python dataset. They contain one or more characters with similar quotation marks. For example:

```
1. >>>single_quote = 'This is an atomic value'
2. >>>double_quote = "Hey it is my watch"
3. >>>single_char_string = "A"
4. >>>empty_string = ""
5. >>>empty_string = ''
6. >>>single_within_double_quote = "hello python
programming language."
7. >>>double_within_single_quote = "Why person is so
'smart'?"
8. >>>same_quotes = 'I\'ve an clue'
9. >>>triple_quote_string = '''This
... is
... triple
... quote'''
10. >>>triple_quote_string
'This\nis\ntriple\nquote'
11. >>>type(single_quote)
<class 'str'>
```

DOI: 10.1201/9781003202035-10

Either single or double quotation marks 2 are embedded in the strings. You just need to add a string to a variable to store a string. In all variables on the left of the assignment operator the code above contains array variables. String 3 is sometimes used as one thread. A series does not need any "actors." The fourth and fifth strings, known as null strings, are all real strings. A double quote string may be a single quote mark 6. Similarly, a string in one quote mark will have double quotes 7. You would use the other form of quotation mark in which the string is followed by one. You must preface the internal quote with the backslash 8 if you want to use the same quotation marks within the text that you have used for the text. When you have a series that covers many lines, it can be found in three quotes 9. The string 10 simply represents both white space and newlines in the three quotations. Bypassing it as an argument into the sort () function, you will find the type of value. Str code set 11 Python strings are type define class.

10.2.1 String str() Function

The str() procedure returns a string assumed to symbolize the object and produce nice output [75]. The str() function syntax is:

str(object)

It returns the object's string form. If the object is not given, a null string will be returned.

```
1. >>>str(20)
       '20'
2. >>>create_string = str()
3. >>>type(creare_string)
       <class 'str'>
```

In this case, the form of an integer is translated to text. Note the individual quotes for series 1. The build string is a string 2 of str 3 sort.

10.3 Basic String Operation

For the development of a repetitive string sequence, Python strings can be paired with the + sign and * operator [76].

```
1.  >>>string_1 = "Insta"
2.  >>>string_2 = "gram"
3.  >>>concatenated_string = string_1 +string_2
4.  >>>concatenated_string
'Instagram'
5.  >>>concatenated_string_with_space = "Hello " +"sir"
6.  >>>concatenated_string_with_space
'Hello sir'
7.  >>>singer = 60 +"cent"
Traceback (most recent call last):
File "<stdin>", line 1, in <module>
TypeError: unsupported operand type(s) for +: 'int' and
'str'
8.  >>>single = str(60) +"cent"
9.  >>>single
'60cent'
10. >>>repetition_of_string = "wow" * 5
11. >>>repetition_of_string
'wowwowwowwowwow'
```

In the "Insta" 1 and "gram" 2 string values, two string variables are allocated. String 1 and string 2 are merged to create a new string with a + operator. The sum of the two strings 3 is the new concatenated string 4. As you can see in the display, the two concatenated string values have no space. All you need to do is to use whitespace in a string as in 5 in restricted strings 6. The + operator cannot be used to merge values of two different forms. For example, the data string type with integer number type 7 cannot be modified. The type integer must be translated to string type, and then values 8–9 must be merged. The multiplication operator * can be used on a number 10. It repeats the number of times you decide for the string, and 5 times 11 times the string name "wow."

With the in and not in membership operator, you can verify the location of a variable on another number. It returns a True or a False Boolean. The operator in True will determine if the string value occurs in the right operand in a character set. The in operator would not equate to True if in the left operand the string value is not represented in the right operand sequence of string characters.

```
1.  >>>fruit_string = "orange is a fruit"
2.  >>>fruit_sub_string = "orange"
3.  >>>fruit_sub_string in fruit_string
True
4.  >>>another_fruit_string = "banana"
5.  >>>another_fruit_string not in fruit_string
True
```

Point 3 is valid since the "orange" string is found in the "orange is a fruit" series. The nonoperative evaluates to True if the string "black" in string 5 is not present.

10.3.1 String Comparison

The two strings that result in either Boolean True or False can be compared using the   >, <, < =, > =, = =, ! = operators: ASCII character worth contrasts strings with Python [77]. For instance:

```
1. >>>"January" == "Janry"
False
2. >>>" January " != " Janry"
True
3. >>>" January " <" Janry"
False
4. >>>" January " >" Janry"
True
5. >>>" January " <= " Janry"
False
6. >>>" January " >= " Janry"
True
7. >>>"filled" >""
True
```

Strings can be contrasted with separate operators of contrast 1–7. Compare string equality with = = (double equal sign). Unfair strings are compared with the ! = symbol. Assume string 1 is called "January" and string 2 is called "Janry". Compare the very first two main characters of string 1 and string 2 (J and J). The two additional characters (a and a) are compared since they are similar. The third two characters (n and n) are related since they are similar too. The fourth character in each string is compared, and since the value for 'u' is larger than the 'e' string, 1 is larger than string 2, as the third characters are both similar. One string can also be compared to an infinite string.

10.3.2 Built-in Functions Used on Strings

Much built-in functionality may be provided by a string as an argument (Table 10.1).

TABLE 10.1

Built-in Functions of a String

Built-in Functions	Description
len()	The len() function calculates the number of characters in a string. The white space characters are also counted.
max()	The max() function returns a character having the highest ASCII value.
min()	The min() function returns a character having the lowest ASCII value.

```
1. >>>count_characters = len("functions")
2. >>>count_characters
9
3. >>>max("square")
'u'
4. >>>min("april")
'a'
```

In the string "functions" 1 the number of characters is calculated using the `len()` function 2. The highest and lowest ASCII characters can be determined using the functions `max()` 3 and `min()` 4.

10.4 Accessing Characters by the Index Number in a String

Each string character is put in the string. The character of each string is the number of the index. The first is index 0. The second character is index 1, etc. The string length is the number of characters in the string [78]. A subscript operator, e.g. a square bracket, will access each character in a series. Place brackets are used to index the value to a certain index or position in a string. This is also regarded as a delivery service. The index breakdown of the string variable "hi beautiful" is:

String	h	i		b	e	a	u	t	i	f	u	l	
	0	1	2	3	4	5	6	7	8	9	10	11	Index

As shown below, the syntax is used to access an individual character in a string:

```
string_name[index]
```

where the index is typically less than that of the string length and between 0 and 1. The index value is always an integer and signifies the character that you want to acquire.

For example:

```
1. >>>String = "hi beautiful"
2. >>> String [0]
'h'
3 >>> String [1]
'i'
4. >>> String [2]
' '
5. >>> String [3]
'b'
6. >>> String [11]
'l'
7. >>> String [12]
Traceback (most recent call last):
File "<stdin>", line 1, in <module>
IndexError: string index out of range
```

You can access each character in a string by referring to index numbers in a square bracket. The index number begins at zero, which in string 2 matches the first character. As we migrate to the next letter, right of the current letter 3–6, the index number increases by one. The spatial character has an index number of its own, i.e. 2. The final character of the strings is the (string size − 1) or (len(string) − 1) index number. If you attempt to put an index number above the number of group of characters in the file, it causes an IndexError: index value out of range 7.

You may also use negative indexing to access individual characters on a chart. If you have a long string and want to add an end character of a string you can count backwards from the bottom of the string, starting from the -1 index number. The negative index divide for the vector string "be yourself" is:

word phrase	b	e		y	o	u	r	s	e	l	f	
	−11	−10	−9	−8	−7	−6	−5	−4	−3	−2	−1	Index

```
1. >>>word_phrase [-1]
'f'
2. >>>word_phrase [-2]
'l'
```

With -1 you can print the character 'f'1, with -2, the character 'l' 2, if you have the negative index number. If you want to enter characters at the end of a long series, you will benefit from negative indexing.

10.5 Slicing and Joining in Strings

The "slice" notation is a convenient way to refer to sub-slices in an initial string sequence [79]. The string slice syntax is:

```
string-name [start:stop[:step]]
```

You can view a string sequence by entering a set of index numbers that are divided by a column and by the slice. The string break returns a series of characters from the start to the end, but without the end. The indexing values must be an integer at the beginning and end. Positive or negative indexing is used for string slicing.

Health_drink	M	I	L	K	Y		T	E	A	
	0	1	2	3	4	5	6	7	8	Index

```
1. >>>Healthy_drink = "MILKY TEA"
2. >>>Healthy_drink[0:3]
'MIL'
3. >>>Healthy_drink[:5]
'MILKY'
4. >>>Healthy_drink[6:]
'TEA'
5. >>>Healthy_drink[:]
'MILKY TEA'
6. >>>Healthy_drink[4:4]
' '
7. >>>Healthy_drink[6:20]
'TEA'
```

A substring, which is a string already present inside another string, is generated when you trim strings. A substring is a string-contained series of instructions. The "green tea" string is allocated to healthy_drinkvariable 1, and from the start of the 0th index to the third (2nd index) a series of characters or substring is extracted 2. In the splitting of the string, the value of the start index indicates where the split begins, and where the splitting ends. If the start index is excluded, the slice begins from the first 0th index to the last index, in three substrates from the 0th to the 4th list display.

The slice starts from the beginning index and runs to the end if the end index is skipped. A substratum is seen in 4 from the sixth to the end of the line. The whole string appears 5 if the index of the beginning and end is omitted. The result is an empty string 6 if the starting index is the same or higher than the end index. If the index value is above the string's end, this can also be achieved at the end of the range 7.

Health_drink	M	I	L	K	Y		T	E	A	
	−9	−8	−7	−6	−5	−4	−3	−2	−1	Index

You may use the negative indexes in a string for specific characters. Negative indexing begins with -1 and fits the last character of a series and, if we step to the left, then the index reduces by one.

```
1. >>>Healthy_drink[-3:-1]
'TE'
2. >>> Healthy_drink[6:-1]
'TE'
```

If you use negative index values as in the earlier series, you must determine the lowest negative integer in the starting index position. Positive and negative index numbers 2 can be merged.

10.5.1 Specifying the Steps of a Slice Operation

A third argument called a phase, that is an alternative, with beginning and finishing index numbers, can be defined in the slicing procedure [80]. This is more like the number that can be missed in the string following the beginning indexing function. The basic phase value is 1. Step numbers are not specified in the previous slice cases and a default value of 1 is used in its absence. For example:

```
1. >>>newspaper = "new york times"
2. >>>newspaper [0:12:4]
'ny'
3. >>>newspaper [::4]
'ny e'
```

In 2 the slice [0:12:4] is set to n and prints each fourth character in a string until the 12th character (excluded) is set to n. You can completely ignore considering all index values in the string by identifying two columns while taking into consideration the step argument, which specifies the set of characters to miss 3.

10.5.2 Joining Strings Using the join() Method

The join() string can add strings. An adaptable way for levels of academic strings is provided by a combination [81]. The join() method syntax is:
string_name.join(sequence)

The string or list series here might be the following. When the series is a string, then join() inserts the string name for each sequence character and returns the merged string. If a list is a chain, then the join() function inserts the string name for any chain object and returns the string confined. Note that the string form should be used for all the items in the list.

```
1. >>>date_of_birth = ["18", "10", "1973"]
2. >>> ":" join(date_of_birth)
'18:19:1973'
3. >>>social_app = ["facebook", "is", "an", "social
media", "sharing", "application"]
4. >>> "".join(social_app)
'facebook is an social media sharing application'
5. >>>numbers = "456"
6. >>>characters = "tim"
7. >>>password = numbers.join(characters)
8. >>>password
't456i456m'
```

The string type is all objects in the date of birth list field. The second string ":" will be added and the string will be shown in each list object. All list items are string form 3 in the social app list field. In 4, the join() method guarantees that a null space for any of the objects in the social app is added, and the merged string is shown. The fifth and sixth variables are of string sort. The "456" string is put for each "amy" string character, resulting in "t456i456m". The value for string "456" is passed from m to m and again from m to you and the password number attribute is allocated.

10.5.3 Split Strings Using the split () Method

A list of string elements is included in the split () method [82]. The process of split () is the code:

string_name.split([separator [, maxsplit]])

The comma-separated string is here separate and available. A single string is separated into a string list depending on the separator defined. White space is known to be the delimiter string to distinguish the strings if the separator is not shown. If maxsplit is given, the maximum splits will be completed (so the full maxsplit +1 items will be in the list). If not defined as maxsplits or −1, then the number of separates is not reduced.

```
1. >>>inventors = "Dennis, goras, james, newton"
2. >>>inventors.split(",")
['edison', 'goras', 'james', 'newton']
3. >>>watches = "rolex timex kapila titan"
4. >>>watches.split()
['rolex', 'timex', 'kapila', 'titan']
```

The value is divided based on the "," (comma) 2 separator in the inventors string vector 1. In 4, the split() method does not indicate any separator. The string variable watches 3 is also whitespace-dependent segregated.

10.5.4 Immutable Strings

Characters in a string cannot be changed when a value for a number attribute is set [83]. However, the same string attribute may be applied to various number values.

```
1. >>>immutable = "dollar"
2. >>>immutable[0] = "c"
Traceback (most recent call last):
File "<stdin>", line 1, in <module>
TypeError: 'str' object does not support item assignment
3. >>>string_immutable = "c" +immut able[1:]
4. >>>string_immutable
'dollar'
5. >>>immutable = "dollar"
6. >>>immutable
'dollar'
```

The "dollar" string is set to the enduring 1 string attribute. If a character is given by indexing, if you attempt to modify the string, the outcome shows as an error since the string is unchanging. 2. You cannot adjust the value of a string until a string attribute has been set. You should render a new string, which is an initial 3-4 line variant. A whole new string attribute is often permitted to be added 5–6 to the current string variable.

10.5.5 String Traversing

As the sequence is a character chain, each symbol can be crossed with the chain [84].

```
def main():
    alphabet = "google"
    index = 0
    print(f"In the string '{alphabet}'")
    for each_character in alphabet:
        print(f"Character '{each_character}' has an index value
        of {index}")
        index += 1
if __name__ == "__main__":
    main()
```

Output

```
In the string 'google'
Character 'g' has an index value of 0
Character 'o' has an index value of 1
Character 'o' has an index value of 2
Character 'g' has an index value of 3
Character 'l' has an index value of 4
Character 'e' has an index value of 5
```

The "internet" string number is set to a variable of the alphabet 2, and the variable of the index to zero 3. It is simple to loop across each symbol in a string with the argument. The index value is raised to one 7 for each character traversed in string 5.

The corresponding index value of each character in the string is written in 6.

10.6 String Methods

The str function is forwarded to dir(); you can obtain a list of all the methods associated with the string from Table 10.2.

TABLE 10.2

Different String Methods

	Syntax	Description
capitalize()	string_name.capitalize()	The capitalize() method returns a copy of the string with its first character capitalized and the rest lower-cased.

(Continued)

TABLE 10.2 (CONTINUED)

	Syntax	Description
casefold()	string_name.casefold()	The casefold() method returns a casefolded copy of the string. Casefolded strings may be used for caseless matching.
center()	string_name.center(width[, fillchar])	The method center() centers string_name by taking the width parameter into account. Padding is specified by parameter fillchar. The default filler is a space.
count()	string_name. count(substring [, start [, end]])	The method count() returns the number of nonoverlapping occurrences of the substring in the range [start, end]. Optional arguments start and end are interpreted as in slice notation.
endswith()	string_name. endswith(suffix[, start[, end]])	This method endswith(), returns True if string_name ends with the specified suffix substring, otherwise it returns False. With option start, the test begins at that position. With option end, comparing stops at that position.
find()	string_name. find(substring[, start[, end]])	Checks if substring appears in string_name or if substring appears in string_name specified by starting index start and ending index end. Returns position of the first character of the first instance of string substring in string_name, otherwise return -1 if substring not found in string_name.
isalnum()	string_name.isalnum()	The method isalnum() returns Boolean True if all characters in the string are alphanumeric and there is at least one character, otherwise it returns Boolean False.
isalpha()	string_name.isalpha()	The method isalpha(), returns Boolean True if all characters in the string are alphabetic and there is at least one character, otherwise it returns Boolean False.
isdecimal()	string_name.isdecimal()	The method isdecimal() returns Boolean True if all characters in the string are decimal characters and there is at least one character, otherwise it returns Boolean False.
isdigit()	string_name.isdigit()	The method isdigit() returns Boolean True if all characters in the string are digits and there is at least one character, otherwise it returns Boolean False.
isidentifier()	string_name.isidentifier()	The method isidentifier() returns Boolean True if the string is a valid identifier, otherwise it returns Boolean False.
islower()	string_name.islower()	The method islower() returns Boolean True if all characters in the string are lowercase, otherwise it returns Boolean False.

(Continued)

TABLE 10.2   (CONTINUED)

	Syntax	Description
isspace()	string_name.isspace()	The method isspace() returns Boolean True if there are only whitespace characters in the string and there is at least one character, otherwise it returns Boolean False.
isnumeric()	string_name.isnumeric()	The method isnumeric(), returns Boolean True if all characters in the string are numeric characters, and there is at least one character, otherwise it returns Boolean False. Numeric characters include digit characters and all characters that have the Unicode numeric value property.
istitle()	string_name.istitle()	The method istitle() returns Boolean True if the string is a title-cased string and there is at least one character, otherwise it returns Boolean False.
isupper()	string_name.isupper()	The method isupper() returns Boolean True if all cased characters in the string are uppercase and there is at least one cased character, otherwise it returns Boolean False.
upper()	string_name.upper()	The method upper() converts lowercase letters in string to uppercase.
lower()	string_name.lower()	The method lower() converts uppercase letters in string to lowercase.
ljust()	string_name.ljust(width[, fillchar])	In the method ljust(), when you provide the string to the method ljust(), it returns the string left justified. The total length of the string is defined in the first parameter of the method width. Padding is done as defined in second parameter fillchar (the default is a space).
rjust()	string_name. rjust(width[,fillchar])	The method rjust() returns the string right justified. The total length of the string is defined in the first parameter of the method width. Padding is done as defined in the second parameter fillchar (the default is a space).
title()	string_name.title()	The method title() returns "titlecased" versions of the string, that is, all words begin with uppercase characters and the rest are lowercase.
swapcase()	string_name.swapcase()	The method swapcase() returns a copy of the string with uppercase characters converted to lowercase, and vice versa.
splitlines()	string_name. splitlines([keepends])	The method splitlines() returns a list of the lines in the string, breaking at line boundaries. Line breaks are not included in the resulting list unless keepends is given and true.

(Continued)

TABLE 10.2   (CONTINUED)

	Syntax	Description
startswith()	string_name. startswith(prefix[, start[, end]])	The method startswith() returns Boolean True if the string starts with the prefix, otherwise return False. Option start tests string_name beginning at that position. Option end, stops comparing string_name at that position.
strip()	string_name.strip([chars])	The method lstrip() returns a copy of string_name in which specified chars have been stripped from both sides of the string. If char is not specified then space is taken as the default.
rstrip()	string_name.rstrip([chars])	The method rstrip() removes all the trailing whitespace of string_name.
lstrip()	string_name.lstrip([chars])	The method lstrip() removes all the leading whitespace in string_name.
replace()	string_name. replace(old, new[, max])	The method replace() replaces all occurrences of old in string_name with new. If the optional argument max is given, then only the first max occurrences are replaced.
zfill()	string_name.zfill(width)	The method zfill() pads string_name with zeros to fill width.

```
1. >>>dir(str)
['__add__', '__class__', '__contains__', '__delattr__',
'__dir__', '__doc__', '__eq__', '__format__', '__ge__',
'__getattribute__', '__getitem__', '__getnewargs__',
'__gt__', '__hash__', '__init__', '__init_subclass__',
'__iter__', '__le__', '__len__', '__lt__', '__mod__',
'__mul__', '__ne__', '__new__', '__reduce__', '__
reduce_ex__', '__repr__', '__rmod__', '__rmul__', '__
setattr__', '__sizeof__', '__str__', '__subclasshook__',
'capitalize', 'casefold', 'center', 'count', 'encode',
'endswith', 'expandtabs', 'find', 'format', 'format_map',
'index', 'isalnum', 'isalpha', 'isdecimal', 'isdigit',
'isidentifier', 'islower',
 'isnumeric', 'isprintable', 'isspace', 'istitle',
'isupper', 'join', 'ljust', 'lower', 'lstrip',
'maketrans', 'partition', 'replace', 'rfind', 'rindex',
'rjust', 'rpartition', 'rsplit', 'rstrip', 'split',
'splitlines', 'startswith', 'strip', 'swapcase', 'title',
'translate', 'upper', 'zfill']
```

For example:

```
1. >>>fact = "Abraham Lincoln was also a champion
wrestler"
2. >>>fact.isalnum()
False
3. >>>"fun".isalpha()
True
4. >>>"2020".isdigit()
True
5. >>>fact.islower()
False
6. >>>"BOMBAY".isupper()
True
7. >>>"rohtak".islower()
True
8. >>>warriors = "aryans knight were vegetarians"
9. >>>warriors.endswith("vegetarians")
True
10. >>>warriors.startswith("aryans")
True
11. >>>warriors.startswith("A")
False
12. >>>warriors.startswith("a")
True
13. >>>"cucumber".find("cu")
0
14. >>>"cucumber".find("um")
3
15. >>>"cucumber".find("xyz")
-1
16. >>>warriors.count("a")
5
17. >>>species = "maxplank father of quantum physics"
18. >>>species.capitalize()
' Maxplank father of quantam physics '
19. >>>species.title()
' Maxplank Father Of Quantam Physics '
20. >>>"Quantam".lower()
' quantam '
21. >>>"quantam".upper()
'QUANTAM '
22. >>>"Centennial Light".swapcase()
'cENTENNIAL lIGHT'
23. >>>"history does repeat".replace("does", "will")
'history will repeat'
24. >>>quote = " Never Stop Dreaming "
```

```
25. >>>quote.rstrip()
' Never Stop Dreaming'
26. >>>quote.lstrip()
'Never Stop Dreaming '
2 7. >>>quote.strip()
'Never Stop Dreaming'
28. >>>'ab c\n\nde fg\rkl\r\n'.splitlines()
['ab c', '', 'de fg', 'kl']
29. >>>"scandinavian countries are rich".center(40)
'scandinavian countries are rich'
```

Different functions are performed by string 1 methods, such as `capital-ise()`. The following are used for translation purposes: `lower()`, `upper()`, `swapcase()`, `title()` and `count()`. The number methods used in string comparisons are `islower()`, `isupper()`, `isdecimal()`, `isdigit()`, `isnumeric()`, `isalpha()` and `isalnum()`. A number of padding string techniques are `rjust()`, `ljust()`, `zfill()` and `centrer()`. To find a sub-string in an established string, the string method `find()` is used. To replace a string, you may use string techniques such as `replace()`, `joint()`, `split()` and `splitlines()`.

10.7 Formatting Strings

There are four major methods, with different strengths and disadvantages, for string formatting in Python [85]. There is a clear rule of thumb for how to choose the right general string formatting solution for your program. It can be very difficult to format strings in Python. Figure 10.1 is a useful chart for doing this.

10.7.1 Format Specifiers

Old

Python strings have a special built-in operation that the percent operator can control [81]. This helps you to quickly customize positions. You will know how this works automatically if you have ever worked with a print-style feature in C. Here is one example:

```
>>> 'Hello, %s' % name
"Hello, Bob"
```

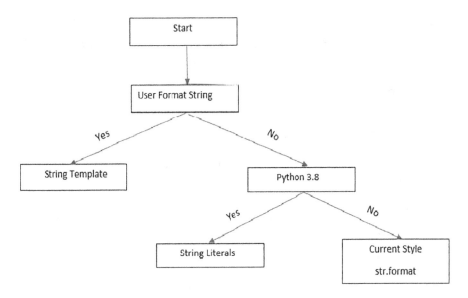

FIGURE 10.1
Writing String Style.

I use the percent %s format specifier to tell Python where the name-value expressed as a string is to be replaced.

New

This "modern" string formatting replaces the peculiar terminology of the percent operator and regularizes string formatting terminology. The encoding of the string object is now performed by calling . (format).

For easy formatting, you can use (format) (just as for formatting in the "classic style"):

```
>>> 'Hello, {}'.format(name)
'Hello, Bob'
```

You can also call and use them in any order you choose for your variable substitutions. This is a very useful function because it allows the view order to be rearranged without modifying the argument.

10.7.2 Escape Sequences

Escape characters are usually used to execute such functions, and their code use allows the programmer to execute a suitable action mapping to that character [79].

For example:

```
'\n'  -->  Leaves a line
'\t'  -->  Leaves a space

# Python code to demonstrate escape character
# string

ch = "I\nLove\tbooks"

print ("The string after resolving escape character is: ")
print (ch)
```

Output

```
The string after resolving the escape character is :
I
Love books
```

But in other situations, no exit is needed, i.e. the entire unfinished string.
Scanning and downloading are accomplished by the following processes.

```
Using repr()
```

In its printable format, this method returns a string that does not address escape sequences.

10.7.3 Raw Strings

A raw string in Python is generated with the prefix 'r' or 'R' [81]. The raw string of Python handles the abstract (\) backslash. This is helpful if we want a backslash string that doesn't have an escape character. This may be helpful.

Python Raw String

Say we're going to build a Hi\nHello string in Python. The \n would be viewed as a new line if we choose to add it to a regular series.

```
s = 'Hi\nHello'
print(s)
```

Output:

```
Hi
Hello
```

See how coarse strings allow one to treat backslash as a natural character.

```
raw_s = r'Hi\nHello'
print(raw_s)
```

Output:

```
Hi\nHello
```

Let's see some cases in which the backslash character has no special significance.

```
>>> s = 'Hi\xHello'
  File "<input>", line 1
SyntaxError: (unicode error)
```

The 'unicodeescape' codec can't decode bytes in position 2–3: truncated \ xXX escape. The error states that Python cannot decipher '\x,' so it has no special significance. Let's see how the same string can be generated with unprocessed strings.

```
>>> s = r'Hi\xHello'
>>> print(s)
Hi\xHello
```

Python Raw Strings and Quotes

If a backslash comes with a pure string quote, then it is an escape. The backslash remains as the outcome, however. We can't build a single back-slash string because of this function [82]. Also, an uncommon number of back-strokes at the end cannot be a raw series.

All raw strings are invalid:

R'\' # absent final quotation when it escapes the final quote

The first two backstrokes are going to escape, the third is going to attempt to escape from the end quote. Let us discuss some relevant instances of raw strings with quotes.

```
raw_s = r'/"
print(raw_s)

raw_s = r'ab\\'
print(raw_s)

raw_s = R'\\\'
" # prefix can be 'R' or 'r'
print(raw_s)
```

Output

```
\'
ab\\
\\\"
```

This is just to add the Python raw string with ease.

10.7.4 Unicodes

Unicode is a major theme. Luckily, to address real-world issues you don't need to know anything about Unicode: a couple of simple bits of information are enough. The distinction between bytes and characters must be understood first. For older languages, bytes and characters in ASCII-centered contexts are the same. For up to 256 values, a byte is limited to 256 characters in these settings. On the other hand, Unicode's characters extend to tens of thousands. This means that each character in Unicode takes about one byte, so you have to differentiate between bytes and characters.

Normal Python strings are simply byte strings, and a character in Python is a letter. The "8-bit string" and "plain string" are standard in Python. We call them byte strings in this tutorial to inform you about their byte orientation.

Conversely, an arbitrary variable large enough to contain a byte, like the long integer in Python, is a Python Unicode script. The representation of Unicode characters becomes a matter of interest only when you attempt to send them to some byte-focused operation such as the write method for files or the sending method for network sockets. You would then determine how the characters are interpreted as bytes. The string is considered to be the encoding of a byte string from Unicode. Likewise, you need to decode strings from bytes to characters when you load Unicode strings from directories, sockets or other byte-oriented objects [76].

There are several ways Unicode artifacts can be translated into byte strings, each called encoding. There is no "correct" coding for a range of geographical, political and technological purposes. Each encoding has its case-insensitive name, which is passed as a parameter to the decoding system:

- Any Unicode character can be used with UTF-8 encoding. It is also back-compatible with ASCII, meaning that a pure ASCII file can both be called a UTF-8 and a UTF-8 file that uses the same characters as an ASCII file. This makes UTF-8 especially with older Unix tools very retro-compatible. The dominant Unix encoding is distant from UTF-8. The biggest flaw is that Eastern alphabets are very unreliable.

- Microsoft OS and the Python system support UTF-16 encoding. For Western languages, it is less efficient but more convenient for Eastern ones. Sometimes a version of UTF-16 is called UCS-2.

- The ASCII 256-character supersets are in the ISO-8859 encoding range. No Unicode character can be supported; they can only support a specific language or language family. In most Western European and African languages, ISO-8859-1 is known as Latin-1, but not Arabic. In Eastern European languages, like Polish and Hungarian, ISO-8859-2, also known as Latin-2, covers many characters.

You probably want to use UTF-8 if you want to encrypt all Unicode characters. You only have to process other encodes when the data from those encodes which are created by another application is transferred to you:

```
# Convert Unicode to plain Python string: "encode"
unicodestring = u"Hello world"
utf8string = unicodestring.encode("utf-8")
asciistring = unicodestring.encode("ascii")
isostring = unicodestring.encode("ISO-8859-1")
utf16string = unicodestring.encode("utf-16")
# Convert plain Python string to Unicode: "decode"
plainstring1 = unicode(utf8string, "utf-8")
plainstring2 = unicode(asciistring, "ascii")
plainstring3 = unicode(isostring, "ISO-8859-1")
plainstring4 = unicode(utf16string, "utf-16")
assert plainstring1==plainstring2=plainstring3==plainstr
ing4
```

10.8 Conclusion

A string is a sequence of characters. To access values through slicing, square brackets are used along with the index. Various string operations include conversion, comparing strings, padding, finding a substring in an existing string and replacing a string in Python. Python strings are immutable which means that once created they cannot be changed.

11

Object-oriented Programming

11.1 Introduction

The basic idea of object-oriented programming (OOP) is that we use objects to model real-world things that we want to represent inside our programs and provide a simple way to access their functionality that would otherwise be hard or impossible to utilize. Large programs are challenging to write. Once a program reaches a certain size, object-oriented programs are actually easier to program than non-object-oriented ones. As programs are not disposable, an object-oriented program is much easier to modify and maintain than a non-object-oriented program. So, object-oriented programs require less work to maintain over time. OOP results in code reuse, cleaner code, better architecture, abstraction layers and fewer programming bugs. Python provides full support for OOP, including encapsulation, inheritance and polymorphism [86].

Overview of OOP terminology:

- **Class**: A user-defined prototype for an object that defines a set of attributes that characterize any object of the class. The attributes are data members (class variables and instance variables) and methods, accessed via dot notation.
- **Class variable**: A variable that is shared by all instances of a class. Class variables are defined within a class but outside any of the class's methods. Class variables aren't used as frequently as instance variables are.
- **Data member**: A class variable or instance variable that holds data associated with a class and its objects.
- **Function overloading**: The assignment of more than one behavior to a particular function. The operation performed varies by the types of objects (arguments) involved.
- **Instance variable**: A variable that is defined inside a method and belongs only to the current instance of a class.

DOI: 10.1201/9781003202035-11

- **Inheritance**: The transfer of the characteristics of a class to other classes that are derived from it.
- **Instance**: An individual object of a certain class. An object obj that belongs to a class Circle, for example, is an instance of the class Circle.
- **Method**: A special kind of function that is defined in a class definition.
- **Object**: A unique instance of a data structure that's defined by its class. An object comprises both data members (class variables and instance variables) and methods.
- **Operator overloading**: The assignment of more than one function to a particular operator.

11.2 Classes and Objects

A class is essentially a logical entity that serves as a blueprint or blueprints to generate objects, whereas an object is a description of Python functions and variables. Included within the class and accessible utilizing objects are variables and attributes. These mechanisms and quantities are known as attributes [87].

Let's take an example for understanding the concept of objects and classes in Python. One object can be thought of as a normal everyday object, for example, a truck. As already explained, we know the data and functions in a class are described and all these documents and records can be viewed as the characteristics and behavior of the object. That is, the car's (object's) features (data) are color, weight, door number and so on. The measurements (features) of the car (object) are rpm, braking and so on. A class, as shown in Figure 11.1, can be used to create multiple objects with different data and associated functions.

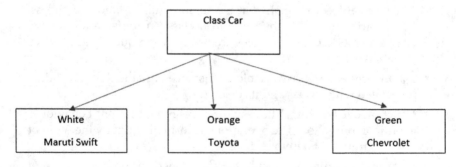

FIGURE 11.1
Class and objects.

Advantages of Using Classes in Python

- Classes offer an easy way to hold data members and strategies in one location that helps to organize the curriculum more effectively.
- Another aspect of this object-oriented paradigm, namely inheritance, is given in classes.
- Classes also help circumvent any default operator.
- The use of classes allows the reuse of the code, which improves the efficiency of the program.
- Description of and maintaining relevant services in one location (classroom) gives a simple code structure that improves software usability.

11.3 Creating Classes in Python

Classes share associated variables and methods [88]. This is the shortest form of the class definition:

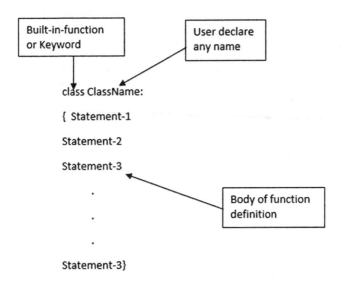

The class statement creates a new class definition. The name of the class immediately follows the keyword class followed by a colon:

```
class ClassName:
    'Optional class documentation string'    class suite
```

- The class has a documentation string, which can be accessed via ClassName.__doc__.
- The class_suite consists of all the component statements defining class members, data attributes and functions.

The following is an example of a simple Python class:

```python
class Employee:
    'Common base class for all employees'
    empCount = 0
    def __init__ (self, name, salary):
        self.name = name        self.salary = salary
        Employee.empCount += 1
        def displayCount(self):
            print ("Total Employee %d" % Employee.
            empCount)
    def displayEmployee(self):
        print ("Name : ", self.name, ", Salary: ", self.
        salary)
```

- The variable empCount is a class variable whose value is shared among all instances of this class. This can be accessed as Employee. empCount from inside the class or outside the class.
- The first method __init__ () is a special method, which is called a class constructor or initialization method that Python calls when you create a new instance of this class.
- You declare other class methods like normal functions with the exception that the first argument in each method is self. Python adds the self argument to the list for you; you don't need to include it when you call the methods.

11.4 How to Create Objects in Python

Creating instance objects
To create instances of a class, you call the class using the class name and pass in whatever arguments its __init__ method accepts.

```
"This would create first object of Employee class" emp1 =
Employee("Zara", 2000)
"This would create second object of Employee class" emp2
= Employee ("Manni", 5000)
```

Accessing attributes:

You access the object's attributes using the dot operator with the object. The class variable is accessed using the class name as follows:

```
emp1.displayEmployee()
emp2.displayEmployee()
print ("Total Employee %d" % Employee.empCount)
```

Now, putting all the concepts together:

```
class Employee:
  'Common base class for all employees'
  empCount = 0

  def __init__(self, name, salary):
    self.name = name
    self.salary = salary
    Employee.empCount += 1

  def displayCount(self):
    print ("Total Employee %d" % Employee.empCount)

  def displayEmployee(self):
    print ("Name : ", self.name,   ", Salary: ", self.salary)

"This would create first object of Employee class"
emp1 = Employee("Zara", 2000)
"This would create second object of Employee class"
emp2 = Employee("Manni", 5000)
emp1.displayEmployee()
emp2.displayEmployee()
print ("Total Employee %d" % Employee.empCount)
```

When the above code is executed, it produces the following result:

```
Name : Zara , Salary: 2000
Name : Manni , Salary: 5000
Total Employee 2
```

You can add, remove or modify attributes of classes and objects at any time:

```
emp1.age = 7  # Add an 'age' attribute.
emp1.age = 8  # Modify 'age' attribute.
del emp1.age  # Delete 'age' attribute.
```

Instead of using the normal statements to access attributes, you can use following functions:

- The **getattr(obj, name[, default])**: to access the attribute of an object.
- The **hasattr(obj,name)**: to check if an attribute exists or not.
- The **setattr(obj,name,value)**: to set an attribute. If the attribute does not exist, then it is created.
- The **delattr(obj, name)**: to delete an attribute.

```
hasattr(emp1, 'age')      # Returns true if 'age'
attribute exists
getattr(emp1, 'age')      # Returns value of 'age'
attribute
setattr(emp1, 'age', 8) # Set attribute 'age' to 8
delattr(emp1, 'age')      # Delete attribute 'age'
```

Built-in class attributes

Every Python class keeps following built-in attributes which can be accessed using the dot operator like any other attribute:

- __**dict**__: Dictionary containing the class's namespace.
- __**doc**__: Class documentation string or None if undefined.
- __**name**__: Class name.
- __**module**__: Module name in which the class is defined. This attribute is "__main__" in interactive mode.
- __**bases**__: A possibly empty tuple containing the base classes, in the order of their occurrence in the base class list.

For the above class let's try to access all these attributes:

```
class Employee:
  'Common base class for all employees'
  empCount = 0

  def __init__(self, name, salary):
    self.name = name
    self.salary = salary
    Employee.empCount += 1
```

```
    def displayCount(self):
     print ("Total Employee %d" % Employee.empCount)

    def displayEmployee(self):
     print ("Name : ", self.name,  ", Salary: ", self.salary)
print ("Employee.__doc__:", Employee.__doc__)
print ("Employee.__name__:", Employee.__name__)
print ("Employee.__module__:", Employee.__module__)
print ("Employee.__bases__:", Employee.__bases__)
print ("Employee.__dict__:", Employee.__dict__)
```

When the above code is executed, it produces the following result:

```
Employee.__doc__: Common base class for all employees
Employee.__name__: Employee
Employee.__module__: __main__
Employee.__bases__: (<class 'object'>,)
Employee.__dict__: {'__module__': ' __main__' '__doc__':
'Common base class for all employees', 'empCount':
0, '__init__': <function Employee.__init__ at
0x0000025F63138DC0>, 'displayEmployee': <function
Employee.displayEmployee at 0x0000025F63138E50>, '__
dict__': <attribute '__dict__' of 'Employee' objects>,
'__weak
ref__': <attribute '__weakref__' of 'Employee' objects>}
```

Destroying objects (garbage collection)

Python deletes unneeded objects (built-in types or class instances) automatically to free memory space. The process by which Python periodically reclaims blocks of memory that no longer are in use is termed "garbage collection."

Python's garbage collector runs during program execution and is triggered when an object's reference count reaches zero. An object's reference count changes as the number of aliases that point to it changes.

An object's reference count increases when it's assigned a new name or placed in a container (list, tuple or dictionary). The object's reference count decreases when it's deleted using del, its reference is reassigned or its reference goes out of scope. When an object's reference count reaches zero, Python collects it automatically.

```
a = 40      # Create object <40> b = a       # Increase ref.
count  of <40>
c = [b]     # Increase ref. count  of <40>
del a       # Decrease ref. count  of <40>
b = 100     # Decrease ref. count  of <40>
c[0] = -1   # Decrease ref. count  of <40>
```

You normally won't notice when the garbage collector destroys an orphaned instance and reclaims its space. But a class can implement the special method __del__(), called a destructor, that is invoked when the instance is about to be destroyed. This method might be used to clean up any nonmemory resources used by an instance.

Example

This __del__() destructor prints the class name of an instance that is about to be destroyed:

```
class Point:
    def __init__( self, x=0, y=0):
        self.x = x
        self.y = y
    def __del__(self):
        class_name = self.__class__.__name__
        print (class_name, "destroyed")

pt1 = Point()
pt2 = pt1
pt3 = pt1
print (id(pt1), id(pt2), id(pt3)) # prints the ids of the
obejcts
del pt1
del pt2
del pt3
```

When the above code is executed, it produces the following result:

```
1650815324016 1650815324016 1650815324016
Point destroyed
```

11.5 The Constructor Method

A constructor is a special type of method (function) which is used to initialize the instance members of a class.

In C++ or Java, the constructor has the same name as its class, but it is treated differently in Python. Here it is used to create an object.

Constructors can be of two types:

1. A parameterized constructor;
2. A non-parameterized constructor.

Constructor definition is executed when we create the object of this class. Constructors also verify that there are enough resources for the object to perform any start-up task.

Creating the constructor in Python

In Python, the method called __init__() simulates the constructor of the class. This method is called when the class is instantiated. It accepts the self keyword as a first argument which allows accessing the attributes or method of the class.

We can pass any number of arguments at the time of creating the class object, depending upon the __init__() definition. It is mostly used to initialize the class attributes. Every class must have a constructor, even if it simply relies on the default constructor.

Consider the following example to initialize the Employee class attributes.

```python
class Employee:

    def __init__(self, i):
        self.employee_id = i

    def work(self):
        print(f'{self.employee_id} is working')

emp = Employee(100)
emp.work()
```

Output

```
100 is working
```

Counting the number of objects of a class

The constructor is called automatically when we create the object of the class. Consider the following example.

```python
class Student:
    count = 0
    def __init__(self):
        Student.count = Student.count + 1

s1=Student()
s2=Student()
s3=Student()
s4=Student()
```

```
s5=Student()
print("The number of student:",Student.count)
```

Output

```
The number of students: 5
```

Python non-parameterized constructor

The non-parameterized constructor is used when we do not want to manipulate the value or when the constructor has only self as an argument. Consider the following example.

```
class Student:
    def __init__(self):
        print("This is non parametrized constructor")
    def show(self,name):
        print("Hello",name)
student = Student()
student.show("python")
```

Output

```
This is non parametrized constructor
Hello python
```

Python parameterized constructor

The parameterized constructor has multiple parameters along with the self. Consider the following example.

```
class Student:
    def __init__(self, name):
        print("This is parametrized constructor")
        self.name = name
    def show(self):
        print("Hello",self.name)
student = Student("python")
student.show()
```

Output

```
This is parametrized constructor
Hello Python
```

Python default constructor

When we do not include the constructor in the class or forget to declare it, then that becomes the default constructor. It does not perform any task but initializes the objects. Consider the following example.

```
class Student:
    roll_num = 101
    name = "Joseph"
    def display(self):
        print(self.roll_num, self.name)
st = Student()
st.display()
```

Output

```
101 Joseph
```

More than one constructor in single class

Let's have a look at another scenario: what happens if we declare the same two constructors in the class.

```
class Student:
    def __init__ (self):
        print("The First Constructor")
    def __init__ (self):
        print("The second contructor")
st = Student()
```

Output

```
The second contructor
```

In the above code, the object st is called the second constructor where both have the same configuration. The first method is not accessible by the st object. Internally, the object of the class will always call the last constructor if the class has multiple constructors.

Constructor overloading is not allowed in Python.

Python built-in class functions

The built-in functions defined in the class are described in Table 11.1.

TABLE 11.1

Built-in Class Functions

SN	Function	Description
1	getattr(obj,name,default)	To access the attribute of the object.
2	setattr(obj, name,value)	To set a particular value to the specific attribute of an object.
3	delattr(obj, name)	To delete a specific attribute.
4	hasattr(obj, name)	Returns true if the object contains some specific attribute.

11.6 Classes with Multiple Objects

You may add one single copy of the data attributes and methods of the class of those classes to construct several classes for a class [91].

Program 11.1 Write a program for the creation of multiple objects for a class.

```
class Birds:
  def __init__(self, bird_name):
    self.bird_name = bird_name
  def flying_birds(self):
   print(f"{self.bird_name} flies above clouds")
  def non_flying_birds(self):
   print(f"{self.bird_name} is the national bird of
   Australia")
def main():
  vulture = Birds("Griffon Vulture")
  crane = Birds("Common Crane")
  emu = Birds("Emu")
  vulture.flying_birds()
  crane.flying_birds()
  emu.non_flying_birds()
if __name__ == "__main__":
  main()
```

Output

```
Griffon Vulture flies above clouds
Common Crane flies above clouds
Emu is the national bird of Australia
```

For Birdsclass 9, three artifacts—vulture, crane and peacock—are produced. All these objects are of the same class, meaning that they all have the same data attribute, but different values. Objects may also have their particular way of operating on the attributes of their results. A solution to an object of class 4-7 is still invoked. The object receives a separate copy of the system attributes and the class is grouped during object instantiation. This means that accurate data and procedures unique to a given object are used. The variable self is originated by a specific object of the class generated throughout initialization 2–3. With the arguments passed to that instance object 9, the parameters of __init__ () function Object() { [native code] } may start.

We now have three objects with separate data attributes. The output "Vulture flies over the clouds", the output "Common Crane flies above the ground" and the output "Peacock is Indian's national bird" will be "vulture flies over naked" in this scenario, and flying birds will be outputed. Please notice that 2 and 3 use a bird's name. Even if they're similar, they're distinct. The header for the functional specification of bird name __init__ () is used as a parameter while the attribute for instance is the bird's name referenced by itself.

11.6.1 Using Objects as Arguments

An object may be converted, if it as an argument, to a calling function [92].

Program 11.8 Write a program to demonstrate the passing of an object, which is an argument, to a function call.

```
class Track:
  def __init__(self, song, artist):
    self.song = song
    self.artist = artist
def print_track_info(vocalist):
  print(f"Song is '{vocalist.song}'")
  print(f"Artist is '{vocalist.artist}'")
singer = Track("The First Time Ever I Saw Your Face",
"Roberta Flack")
print_track_info(singer)
```

Output

```
Song is "The First Time Ever I Saw Your Face"
Artist is "Roberta Flack"
```

With the class Volume, 1 the song and artist data attributes 2–4 are applied to the __init__ () process. The function print track info() receives an object as parameters 5–7. The Volume class object singer 8 is transferred to print track info() function 9 as a statement.

11.6.2 Objects as Return Values

It should be remembered that the whole of Python, including classes, is an entity [93]. In Python, everything is an object since it has no basic, unboxed values. Anything can be used as a value (int, str, float, method, node, etc.) as an entity. The identifications of the object position in the memory are calculated by the id() function. The id() function syntax is:

$$id(object)$$

The function will return an element's "identity." This is an integer (or long integer) that guarantees that this entity will be special and consistent throughout its lifetime. The id() value can be equivalent to two non-overlapping entities. If an object is an instance or not of a given class, the instance() function is used. Instance (object, class info) is the function syntax of instance(); the class details can be either a list or tuple containing a class or other tuple if the object is an object instance and class data. The instance() function returns an instance or subclass of another object, indicated by a Boolean.

11.7 Difference Between Class Attributes and Data Attributes

Datasets are, generally speaking, instance values that are exclusive to a class object, and entity properties are individual variables that are exchanged with all class objects [94].

Program 11.12 Write a program for class variables and instance variables.

```
class Dog:
  kind = 'canine'
  def __init__(self, name):
    self.dog_name = name
d = Dog('Fido')
e = Dog('Buddy')
print(f"Value for Shared Variable or Class Variable
'kind' is '{d.kind}'")
print(f"Value for Shared Variable or Class Variable
'kind' is '{e.kind}'")
print(f"Value for Unique Variable or Instance Variable
'dog_name' is '{d.dog_name}'")
print(f"Value for Unique Variable or Instance Variable
'dog_name' is '{e.dog_name}'")
```

Output

```
Value for Shared Variable or Class Variable 'kind' is 'canine'
Value for Shared Variable or Class Variable 'kind' is 'canine'
Value for Unique Variable or Instance Variable 'dog_name' is
'Fido'
Value for Unique Variable or Instance Variable 'dog_name' is
'Buddy'
```

The variable type here belongs to a base classifier used by all objects 2. The cat name instance variable is special to every structure supported. We use dot notation to control all class and instance variables. The Cat objects d and e share the same vector form of the class. The `self` parameter is then replaced by the generated object, for example, the vector dog name. The dog name example variable is special to the d and e objects that have been generated to print separate values for each of these objects.

11.8 Encapsulation

The encapsulation definition is used by all programs in this chapter. Encapsulation is one of the main principles of object-driven programming. Encapsulation is the mechanism by which data and methods are merged to form a structure known as a class [95]. The variables are not reached explicitly in the encapsulation; they are accessed by the methods present in the class. Encapsulation means that the inner portrayal of the entity (its status and behavior) is shielded from the rest of the components. The principle of data protection is thus possible with encapsulation. To explain the principle of data protection, if any programmer accesses data stored on the outside of the class variable then it is particularly dangerous if they write their code for managing the data contained in a variable (not encapsulated). This would lead to code replication, at least if a deployment is not completely compliant (e.g. needless efforts), and inconsistency. Data shielding also ensures that everybody uses the methods given to view the data contained in a variable such that they are the same for everyone. A class program doesn't have to specify how it works internally or how it is applied. The software creates an object easily and uses it to call the class methods. Abstraction is a mechanism in which only specific variables are presented and are used for accessing information and "hiding" user-focused object implementation data.

Taking your cellphone into consideration (Figure 11.2), you only need to know which keys to email a message or call. How your message is sent out and how your calls are linked is all abstracted from users by clicking a button.

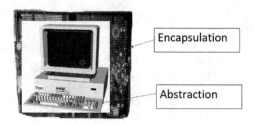

FIGURE 11.2
Abstraction and encapsulation.

Program 11.13. Write a program to show the difference between abstraction and encapsulation.

```
1. class red:
2. def __init__(self, x, y):
3. self.x = x
4. self.y = y
5. def add(self):
6. return self.x +self.y
7. pink_object = red(5.6)
8. red_object.add()
```

The internal image of an object in the red class 1–6 is inaccessible beyond the scope of encapsulation in the above program. Any usable member or even an element red (data/method) is limited and only objects 7–8 can access it. The mechanism is secret ~Abstraction (implementation of add mechanism).

11.8.1 Using Private Instance Variables and Methods

Reference variables or methods available within one class and not available beyond are called variables in a private instance or in private methods [90]. Since there is a legitimate case for using private class members only (in some words to prevent name conflicts with subclass-defined names), such a method, known as name mangling, is acceptable. In Python, an identifier should be considered private (whether it is the function or vector of the instance) when prefixed by two underlines (e.g. spam) and with no sublines. Each type of spam recognition is textually substituted by the class name spam, which has been omitted by the class name of the last name with a large underscore(s). These mangles take place without recognition of the identifier's syntactic role, as long as they fall under the class description. Name manipulation helps encourage subclasses to bypass methods without intraclass naming. Name manipulation aims to provide a means of defining "private" instance variables and methods for classes without caring about the case variables specified by derived classes. Remember, that perhaps the rules for mangling are primarily meant to deter accidents; a value that is called private can also be manipulated or changed.

Program 11.14. Write a program to make private instance variables

```
class PrivateDemo:
  def __init__(self):
    self.nonprivateinstance = "I'm not a private instance"
    self.__privateinstance = "I'm private instance"
  def display_privateinstance(self):
    print(f"{self.__privateinstance} used within the
    method of a class")
def main():
  demo = PrivateDemo()
  print("Invoke Method having private instance")
  print(demo.display_privateinstance())
  print("Invoke non-private instance variable")
  print(demo.nonprivateinstance)
  print("Get attributes of the object")
  print(demo.__dict__)
  print("Trying to access private instance variable
  outside the class results in an error")
  print(demo.__privateinstance)
if __name__ == "__main__":
  main()
```

Output

```
Invoke Method having private instance
I'm private instance used within the method of a class
None
Invoke non-private instance variable
I'm not a private instance
Get attributes of the object
{'nonprivateinstance': "I'm not a private instance",
'_PrivateDemo__privateinstance': "I'm private instance"}
```

The PD object has no attribute 'private instance' in class Private Example 1; Privileged instance 4 cannot be used in the class but can be included in me. {"non-private instance"} ['No Privileged instance': "I'm not private instance']: 'PD' object is not an attribute of 'private instance'. If you are attempting to manipulate private variables outside the class, it could lead to an error. To collect all an object's attributes, you use dict. The private instance variable is prefixed with PrivateDemo, as you can see in the data.

11.9 Inheritance

Inheritance is an important aspect of the object-oriented paradigm. Inheritance provides code reusability to the program so we can then use an existing class to create a new class instead of creating it from scratch.

In inheritance, the child class acquires the properties and can access all the data members and functions defined in the parent class. A child class can also provide its specific implementation to the functions of the parent class. In this section, we will discuss inheritance in detail.

In Python, a derived class can inherit a base class by just mentioning the base in the bracket after the derived class name. Consider the following syntax for inheriting a base class into the derived class:

```
class derived-class(base class):
    <class-suite>
```

A class can inherit multiple classes by mentioning all of them inside the bracket. Consider the following syntax:

```
class derive-class(<base class 1>, <base class 2>, ..... <base
class n>):
    <class - suite>
```

Example

```
class Animal:
  def speak(self):
    print("Animal Speaking")
#child class Dog inherits the base class Animal
class Dog(Animal):
  def bark(self):
    print("dog barking")
d = Dog()
d.bark()
d.speak()
```

Output

```
dog barking
Animal Speaking
```

Python multi-level inheritance

Multi-level inheritance is possible in Python like other object-oriented languages. Multi-level inheritance is archived when a derived class inherits another derived class. There is no limit on the number of levels up to which the multi-level inheritance is archived.

The syntax of multi-level inheritance is:

```
class class1:
    <class-suite>
class class2(class1):
    <class suite>
class class3(class2):
    <class suite>
```

Example

```
class Animal:
    def speak(self):
        print("Animal Speaking")
#The child class Dog inherits the base class Animal
class Dog(Animal):
    def bark(self):
        print("dog barking")
#The child class Dogchild inherits another child class Dog
class DogChild(Dog):
    def eat(self):
        print("Eating bread...")
d = DogChild()
d.bark()
d.speak()
d.eat()
```

Output

```
dog barking
Animal Speaking
Eating bread...
```

Python multiple inheritance

Python provides us with the flexibility to inherit multiple base classes in the child class.

The syntax to perform multiple inheritance is:

```
class Base1:
   <class-suite>

class Base2:
   <class-suite>
.
.
.
class BaseN:
   <class-suite>

class Derived(Base1, Base2, ...... BaseN):
   <class-suite>
```

Example

```
class Calculation1:
   def Summation(self,a,b):
      return a+b;
class Calculation2:
   def Multiplication(self,a,b):
      return a*b;
class Derived(Calculation1,Calculation2):
   def Divide(self,a,b):
      return a/b;
d = Derived()
print(d.Summation(10,20))
print(d.Multiplication(10,20))
print(d.Divide(10,20))
```

Output

```
30
200
0.5
```

The issubclass(sub,sup) method

The issubclass(sub, sup) method is used to check the relationships between the specified classes. It returns true if the first class is the subclass of the second class, and false otherwise.

Consider the following example.

```
class Calculation1:
    def Summation(self,a,b):
        return a+b;
  class Calculation2:
def Multiplication(self,a,b):
        return a*b;
  class Derived(Calculation1,Calculation2):
    def Divide(self,a,b):
        return a/b;
d = Derived()
print(issubclass(Derived,Calculation2))
print(issubclass(Calculation1,Calculation2))
```

Output

```
True
False
```

The isinstance (obj, class) method

The `isinstance()` method is used to check the relationship between the objects and classes. It returns true if the first parameter, i.e. obj, is the instance of the second parameter, i.e. class.

Consider the following example:

```
class Calculation1:
  def Summation(self,a,b):
    return a+b;
class Calculation2:
  def Multiplication(self,a,b):
    return a*b;
class Derived(Calculation1,Calculation2):
  def Divide(self,a,b):
    return a/b;
d = Derived()
print(isinstance(d,Derived))
```

Output

```
True
```

Method verriding

We can provide specific implementation of the parent class method in our child class. When the parent class method is defined in the child class with some specific implementation, then the concept is called method overriding. We may need to perform method overriding in the scenario where a different definition of a parent class method is needed in the child class.

Consider the following example to perform method overriding:

```
class Animal:
  def speak(self):
    print("speaking")
class Dog(Animal):
  def speak(self):
    print("Barking")
d = Dog()
d.speak()
```

Output

```
Barking
```

Real life example of method overriding

```
class Bank:
    def getroi(self):
        return 10;
class SBI(Bank):
    def getroi(self):
        return 7;

class ICICI(Bank):
    def getroi(self):
        return 8;
b1 = Bank()
b2 = SBI()
b3 = ICICI()
print("Bank Rate of interest:",b1.getroi());
print("SBI Rate of interest:",b2.getroi());
print("ICICI Rate of interest:",b3.getroi());
```

Output

```
Bank Rate of interest: 10
SBI Rate of interest: 7
```

```
ICICI Rate of interest: 8
```

Data abstraction in Python

Abstraction is an important aspect of OOP. In Python, we can also perform data hiding by adding the double underscore (___) as a prefix to the attribute which is to be hidden. After this, the attribute will not be visible outside of the class through the object.

Consider the following example:

```
class Employee:
    __count = 0;
    def __init__(self):
      Employee.__count = Employee.__count+1
    def display(self):
        print("The number of employees",Employee.__count)
emp = Employee()
emp2 = Employee()
try:
    print(emp.__count)
finally:
    emp.display()
```

Output

```
The number of employees 2
```

11.10 Polymorphism

"Poly" involves "a variety of forms" and "morphism."

The literal meaning of polymorphism is the condition of occurrence in different forms.

Polymorphism is a very important concept in programming. It refers to the use of a single type entity (method, operator or object) to represent different types in different scenarios.

Let's take an example.

Similarly, for string data types, the + operator is used to perform concatenation.

```
str1 = "Python"
```

```
str2 = "Programming"
```

Example 11.1   Polymorphism in the addition operator

We know that the + operator is used extensively in Python programs. But, it does not have a single usage.

For integer data types, the + operator is used to perform arithmetic addition operation.

```
num1 = 1
num2 = 2
print(num1+num2)
```

Hence, the above program outputs "3".

```
print(str1+" "+str2)
```

As a result, the above program outputs "Python Programming".

Here, we can see that a single operator + has been used to carry out different operations for distinct data types. This is one of the most simple occurrences of polymorphism in Python.

Function polymorphism in Python

There are some functions in Python which are compatible to be run with multiple data types.

One such function is the len() function, which can run with many data types in Python. Let's look at some example-use cases of the function.

Example 11.2   Polymorphic len() function

```
print(len("Programiz"))
print(len(["Python", "Java", "C"]))
print(len({"Name": "John", "Address": "Nepal"}))
```

Output

```
9
3
2
```

Here, we can see that many data types such as string, list, tuple, set and dictionary can work with the len() function. However, we can see that it returns specific information about specific data types.

This is polymorphism in the len() function in Python.

Class polymorphism in Python

Polymorphism is a very important concept in OOP.

We can use it while creating class methods as Python allows different classes to have methods with the same name.

We can then later generalize calling these methods by disregarding the object we are working with. Let's look at an example.

Example 11.3 Polymorphism in class methods

```python
class Cat:
  def__init__(self, name, age):
    self.name = name
    self.age = age

  def info(self):
     print(f"I am a cat. My name is {self.name}. I am
     {self.age} years old.")
  def make_sound(self):
    print("Meow")

class Dog:
  def __init__(self, name, age):
    self.name = name
    self.age = age

  def info(self):
    print(f"I am a dog. My name is {self.name}. I am
    {self.age} years old.")

  def make_sound(self):
    print("Bark")
cat1 = Cat("Kitty", 2.5)
dog1 = Dog("Fluffy", 4)

for animal in (cat1, dog1):
  animal.make_sound()
  animal.info()
  animal.make_sound()
```

Output

```
Meow
I am a cat. My name is Kitty. I am 2.5 years old.
Meow
Bark
I am a dog. My name is Fluffy. I am 4 years old.
Bark
```

Here, we have created two classes `Cat` and `Dog`. They share a similar structure and have the same method names: `info()` and `make_sound()`.

However, notice that we have not created a common superclass or linked the classes together in any way. Even then, we can pack these two different objects into a tuple and iterate through it using a common `animal` variable. This is possible due to polymorphism.

Python method overriding

As in other programming languages, the child classes in Python also inherit methods and attributes from the parent class. We can redefine certain methods and attributes specifically to fit the child class, which is known as "method overriding."

Polymorphism allows us to access these overridden methods and attributes that have the same name as the parent class.

Let's look at an example.

Example 11.4   Method overriding

```python
from math import pi

class Shape:
    def __init__(self, name):
        self.name = name

    def area(self):
        pass

    def fact(self):
        return "I am a two-dimensional shape."

    def __str__(self):
        return self.name

class Square(Shape):
    def __init__(self, length):
        super().__init__("Square")
```

```
            self.length = length
      def area(self):
            return self.length**2

      def fact(self):
            return "Squares have each angle equal to 90
            degrees."
class Circle(Shape):
      def __init__(self, radius):
            super().__init__("Circle")
            self.radius = radius

      def area(self):
            return pi*self.radius**2

a = Square(4)
b = Circle(7)
print(b)
print(b.fact())
print(a.fact())
print(b.area())
```

Output

```
Circle
I am a two-dimensional shape.
Squares have each angle equal to 90 degrees.
153.93804002589985
```

Here, we can see that methods, such as __str__(), which have not been overridden in the child classes, are used from the parent class.

Due to polymorphism, the Python interpreter automatically recognizes that the fact() method for object a(*Square* class) is overridden. So, it uses the one defined in the child class.

On the other hand, since the fact() method for object b isn't overridden, it is used from the parent Shape class.

11.10.1 Python Operator Overloading

Operator overloading is a unique circumstance of polymorphism in which several operators, according to their arguments, have several functions [89]. Python operators work for built-in classes. But the same operator behaves differently with different types. For example, the + operator will perform arithmetic addition on two numbers, merge two lists or concatenate two strings.

This feature that allows the same operator to have different meaning according to the context is called operator overloading.

The class will apply such syntax-invoked operations by setting methods with names known as "magic methods" (Table 11.2), such as arithmetic operations or subscribing and slicing. This is the operator overload approach which is allowed by Python.

TABLE 11.2

Different Operators and Functions

Binary Operators		
Operator	Method	Description
+	__add__(self, other)	For addition operations
-	__sub__(self, other)	Invoked for Subtraction Operations
*	__mul__(self, other)	Invoked for Multiplication Operations
/	__truediv__(self, other)	Invoked for Division Operations
//	__floordiv__(self, other)	Invoked for Floor Division Operations
%	__mod__(self, other)	Invoked for Modulus Operations
**	__pow__(self, other[, modulo])	Invoked for Power Operations
<<	__lshift__(self, other)	Invoked for Left-Shift Operations
>>	__rshift__(self, other)	Invoked for Right-Shift Operations
&	__and__(self, other)	For binary AND operations
^	__xor__(self, other)	Invoked for Binary Exclusive-OR Operations
|	__or__(self, other) Invoked for Binary OR Operations	
Extended Operators		
Operator	Method	Description
+=	_iadd__(self, other)	Invoked for Addition Assignment Operations
-=	__isub(self, other)	Invoked for Subtraction Assignment Operations
*=	__imul__(self, other)	Invoked for Multiplication Assignment Operations
/=	__idiv__(self, other)	Invoked for Division Assignment Operations
//=	__ifloordiv__(self, other)	Invoked for Floor Division Assignment Operations
%=	__imod__(self, other)	Invoked for Modulus Assignment Operations
**=	__ipow__(self, other[, modulo])	Invoked for Power Assignment Operations
<<=	__ilshift__(self, other)	Invoked for Left-Shift Assignment Operations
>>=	__irshift__(self, other)	Invoked for Right-Shift Assignment Operations
&=	__iand__(self, other)	Invoked for Binary AND Assignment Operations
^=	__ixor__(self, other)	Invoked for Binary Exclusive-OR Assignment Operations
|=	__ior__(self, other)	Invoked for Binary OR Assignment Operations

(Continued)

TABLE 11.2 (CONTINUED)

	Unary Operators	
Operator	**Method**	**Description**
-	__neg__(self)	Invoked for Unary Negation Operator
+	__pos__(self)	Invoked for Unary Plus Operator
abs()	__abs__()	Invoked for built-in function abs(). Returns absolute value
~	__invert__(self)	Invoked for Unary Invert Operator

	Conversion Operators	
Functions	**Method**	**Description**
complex()	__complex__(self)	Invoked for built-in complex() function
int()	__int__(self)	Invoked for built-in int() function
long()	__long__(self)	Invoked for built-in long() function
float()	__float__(self)	Invoked for built-in float() function
oct()	__oct__()	Invoked for built-in oct() function
hex()	__hex__()	Invoked for built-in hex() function

	Comparison Operators	
Operator	**Method**	**Description**
<	__lt__(self, other)	Invoked for Less-Than Operations
<=	__le__(self, other)	Invoked for Less-Than or Equal-To Operations
==	__eq__(self, other)	Invoked for Equality Operations
!=	__ne__(self, other)	Invoked for Inequality Operations
>=	__ge__(self, other)	Invoked for Greater Than or Equal-To Operations
>	__gt__(self, other)	Invoked for Greater Than Operations
^=	__ixor__(self, other)	Invoked for Binary Exclusive-OR Assignment Operations
|=	__ior__(self, other)	Invoked for Binary OR Assignment Operations

	Unary Operators	
Operator	**Method**	**Description**
-	__neg__(self)	Invoked for Unary Negation Operator
+	__pos__(self)	Invoked for Unary Plus Operator
abs()	__abs__()	Invoked for built-in function abs(). Returns absolute value
~	__invert__(self)	Invoked for Unary Invert Operator

	Conversion Operators	
Functions	**Method**	**Description**
complex()	__complex__(self)	Invoked for built-in complex() function
int()	__int__(self)	Invoked for built-in int() function
long()	__long__(self)	Invoked for built-in long() function
float()	__float__(self)	Invoked for built-in float() function
oct()	__oct__()	Invoked for built-in oct() function
hex()	__hex__()	Invoked for built-in hex() function

(Continued)

TABLE 11.2   (CONTINUED)

	Comparison Operators	
Operator	Method	Description
<	__lt__(self, other)	Invoked for Less-Than Operations
<=	__le__(self, other)	Invoked for Less-Than or Equal-To Operations
==	__eq__(self, other)	Invoked for Equality Operations
!=	__ne__(self, other)	Invoked for Inequality Operations
>=	__ge__(self, other)	Invoked for Greater Than or Equal-To Operations
>	__gt__(self, other)	Invoked for Greater Than Operations

So what happens when we use them with objects of a user-defined class? Let us consider the following class, which tries to simulate a point in a 2D coordinate system.

```
class Point:
   def __init__(self, x=0, y=0):
      self.x = x
      self.y = y
p1 = Point(1, 2)
p2 = Point(2, 3)
print(p1+p2)
```

Output

```
Traceback (most recent call last):
  File "<string>", line 9, in <module>
    print(p1+p2)
TypeError: unsupported operand type(s) for +: 'Point' and
'Point'
```

Here, we can see that a `TypeError` was raised, since Python didn't know how to add two `Point` objects together.

However, we can achieve this task in Python through operator overloading. But first, let's get some idea about special functions.

Python special functions

Class functions that begin with a double underscore __ are called special functions in Python.

These functions are not the typical functions that we define for a class. The __init__() function we defined above is one of them. It gets called every time we create a new object of that class.

There are numerous other special functions in Python. Visit Python Special Functions to learn more about them.

Using special functions, we can make our class compatible with built-in functions.

```
>>> p1 = Point(2,3)
>>> print(p1)
<__main__.Point object at 0x00000000031F8CC0>
```

Suppose we want the `print()` function to print the coordinates of the `Point` object instead of what we have. We can define a `__str__()` method in our class that controls how the object gets printed. Let's look at how we can achieve this:

```
class Point:
    def __init__(self, x = 0, y = 0):
        self.x = x
        self.y = y

    def __str__(self):
        return "({0},{1})".format(self.x,self.y)
#Now let's try the print() function again.
class Point:
    def __init__(self, x=0, y=0):
        self.x = x
        self.y = y

    def __str__(self):
        return "({0}, {1})".format(self.x, self.y)
p1 = Point(2, 3)
print(p1)
```

Output

```
(2, 3)
```

That's better. Turns out that this same method is invoked when we use the built-in function `str()` or `format()`.

```
>>> str(p1)
'(2,3)'

>>> format(p1)
'(2,3)'
```

So, when you use `str(p1)` or `format(p1)`, Python internally calls the `p1.__str__()` method. Hence the name: special functions.
   Now let's go back to operator overloading.

Overloading the + operator

To overload the + operator, we will need to implement the __add__() function in the class. With great power comes great responsibility. We can do whatever we like, inside this function. But it is more sensible to return a Point object of the coordinate sum.

```
class Point:
    def __init__(self, x=0, y=0):
        self.x = x
        self.y = y

    def __str__(self):
        return "({0},{1})".format(self.x, self.y)

    def __add__(self, other):
        x = self.x + other.x
        y = self.y + other.y
        return Point(x, y)

p1 = Point(1, 2)
p2 = Point(2, 3)

print(p1+p2)
```

Output

```
(3,5)
```

What actually happens is that when you use p1 + p2, Python calls p1.__add__(p2) which in turn is Point.__add__(p1,p2). After this, the addition operation is carried out the way we specified.

Similarly, we can overload other operators as well.

Overloading comparison operators

Python does not limit operator overloading to arithmetic operators only. We can overload comparison operators as well.

Suppose we wanted to implement the less than symbol (<) in our Point class.

Let us compare the magnitude of these points from the origin and return the result for this purpose. This can be implemented as follows.

```
# overloading the less than operator
class Point:
    def __init__(self, x=0, y=0):
        self.x = x
        self.y = y
```

```
    def __str__(self):
        return "({0},{1})".format(self.x, self.y)

    def __lt__(self, other):
        self_mag = (self.x ** 2) + (self.y ** 2)
        other_mag = (other.x ** 2) + (other.y ** 2)
        return self_mag < other_mag

p1 = Point(1,1)
p2 = Point(-2,-3)
p3 = Point(1,-1)

# use less than
print(p1<p2)
print(p2<p3)
print(p1<p3)
```

Output

```
True
False
False
```

11.11 Conclusion

Objects are used to model real-world entities that are represented inside programs; an object is an instance of a class. A class is a blueprint from which individual objects are created. An object is a bundle of related variables and methods. The act of creating an object from a class is called instantiation. The __init__() method is automatically called and executed when an object of the class is created. Class attributes are shared by all the objects of a class. An identifier prefixed with a double underscore and with no trailing underscores should be treated as private within the same class. Encapsulation is the process of combining variables that store data and methods that work on those variables into a single unit called a class. Inheritance enables new classes to receive or inherit variables and methods of existing classes and helps to reuse code.

12

GUI Programming Using Tkinter

12.1 Introduction

Tkinter enables you to design interface programs and is an ideal tool for learning object-driven programming. A lot of interface modules are available in Python for GUI development. The turtle module has been used for the construction of geometric forms. Turtle is a simple to use platform for learning basic programming. Still, you can't make graphical user interfaces (GUIs) from turtle. Tkinter is a valuable tool to develop GUI projects, as well as a precious pedagogical weapon to learn object-oriented programming. Tkinter is short for "Tk Interface," a GUI library used for the creation of Windows, Mac and UNIX GUI programs with many programming languages. Tkinter provides a Python programmer interface. A basic de facto library in Tk GUI is used to construct Python GUI programs [96].

12.2 Getting Started with Tkinter

The Tkinter module requires interface groups. The class Tk can build a window to display GUI (visual components) widgets. Tkinter is actually an inbuilt Python module used to create simple GUI apps. It is the most commonly used module for GUI apps in Python.

You don't need to worry about the installation of the Tkinter module as it comes with Python by default.

We will use the Python 3.6 version for this tutorial. So, kindly update Python if you're using other versions.

Do not blindly copy the code. Try to write by modifying it as you like and then observing and understanding the resulting changes.

Now let us check out the fundamentals of Tkinter so that we can go about creating our own GUIs. To start with, we first import the Tkinter model. Following that, we create the main window. It is in this window that we

DOI: 10.1201/9781003202035-12

perform operations and display visuals, and everything basically. Later, we add the widgets and lastly we enter the main event loop.

There are two keywords here that you might not know at this point:

- Widgets;
- The main event loop.

An event loop is basically telling the code to keep displaying the window until we manually close it. It runs in an infinite loop in the back-end.

```
from tkinter import *
window = Tk()
window.title("Welcome to Tkinter")
lbl = Label(window, text="Hello GUI Programming")
lbl.grid(column=0, row=0)
window.mainloop()
```

FIGURE 12.1
Simple TK example.

As you can see, we import the Tkinter package and define a window. Following that, we provide a window title which is shown on the title tab whenever you open an application.

For example, "Microsoft Word" is shown on the title tab when you open a word application. Similarly here we call it "GUI." We can call it anything we want based on the requirement.

Lastly, we have a label. A label is nothing but what output needs to be shown in the window. In this case, as you can already see, it is "Hello GUI Programming".

Place five lines in the Python command-line interface except one which is the library part of Python (or place in the file "python filename.py") in a Tkinter browser.

Code comments

Initially, we import the Tkinter module [99]. By using the Import Mark tag, the main loop stops linking the tag to the Kit attributes and processes. We build a mark with two lines of text and design the widget using the geometry manager. Finally, we call the principal Tkinter loop to navigate and enable the display. This example does not apply to any app-specific occurrence but a control key loop is essential to display it. Now it can't get any better! The programming of the Tkinter interface is guided by the situation. The software is waiting for user inputs such as mouse clicks and key presses after the user interface has been presented. The accompanying document states this. The declaration establishes a case chain. The event loop constantly analyses the events unit. As seen in Figure 12.2, you close the key window.

The Tkinter hierarchical is, in truth, not a group, unlike certain windowing schemes. A widget class is composed of Widget Management (WM), Misc, Set, Position and Grid. For regular activities, the most a programmer needs to know is the lowest level in the tree; the highest level can be overlooked. Figure 12.3 illustrates the definition of "hierarchical structure."

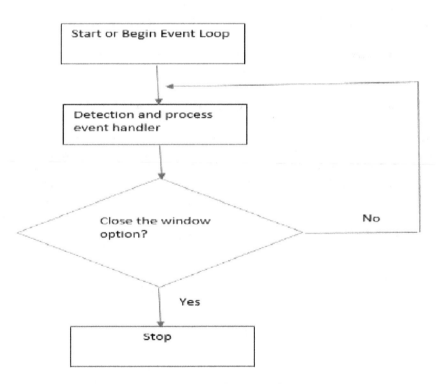

FIGURE 12.2
Flowchart of Tkinter.

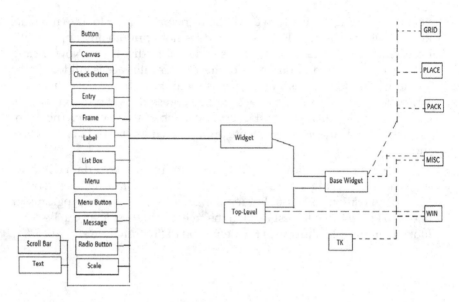

FIGURE 12.3
Hierarchy of Tkinter widget.

12.3 Processing Events

A Tkinter widget may be added to an event called a function [100]. The button slider is a nice way to illustrate simple programming, so we'll use it in the example below. You can use this event if the user clicks on a bell. The Scale widget will set linear values between specified values below and above and graphically shows the current value. The integer value will also be seen. The Scale widget has several different choices, otherwise it is a pretty basic widget. The following is an approximation from one of the examples given with the Tk release, as shown in Figure 12.4. It can also be beneficial for Tk programmers to see how to migrate to Tkinter.

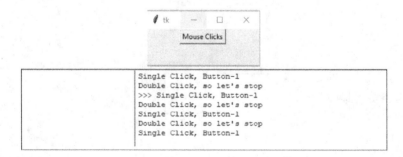

FIGURE 12.4
Processing event.

```
from tkinter import*
def hello (event):
    print("Single Click, Button-1")
def quit(event):
    print("Double Click, so let's stop")
    import sys; sys.exit()

widget = Button(None, text='Mouse  Clicks')
widget.pack()
widget.bind('<Button-1>', hello)
widget.bind('<Double-1>', quit)
widget.mainloop()
```

12.4 The Widget Classes

The following feature demonstrates the standard presentation and use of Tkinter widgets. The code is rather brief and displays just a couple of the widget choices on offer. Often one or more methods of a widget are used, but this scratches only the surface. If you need to look at a specific process or alternative, the corresponding one is accessed by each widget as well. With the exception of the first example, the boiler platform code used to import and initialize Tkinter has been omitted from the code samples. The constant code is shown in bold. Notice that most cases were coded rather than listed as functions. The amount of code, therefore, remains limited. The entire source code is available online for all displays. Widgets are something like elements in HTML. You will find different types of widgets to the elements in Tkinter.

Let's have a brief introduction to all of these widgets in Tkinter.

Here is a list of the main Tkinter widgets:

- Canvas: used to draw shapes in your GUI.
- Button: used to place the buttons in Tkinter.
- Checkbutton: used to create the check buttons in your application. Note that you can select more than one option at a time.
- Entry: widget used to create input fields in the GUI.
- Frame: used as container in Tkinter.
- Label: used to create a single line of widgets like text and images.
- Menu: used to create menus in the GUI.

These widgets are the reason that Tkinter is so popular. It makes it really easy to understand and use practically [101].

12.4.1 Toplevel

For other widgets, like a window, the Toplevel plugin has a different compartment. The rooftop level generated when you initialize Tk can be the only shell you need for simple single-window applications. In Figure 12.5, there are four types of highest level:

- The principal top-level, generally known as the root.
- A child class, which exists separately from the root even if the root is lost.
- A temporary peak, often drawn above the parent and obscured whether the parent is iconified or deleted.
- An undecorated Toplevel can be generated by setting the overridden flag to a null value. This produces an unrealized or moveable frame.

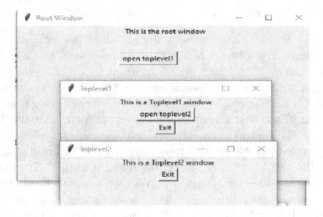

FIGURE 12.5
Top level widgets.

```
from tkinter import *

# Create the root window
# with specified size and title
root = Tk()
root.title("Root Window")
root.geometry("450x300")

# Create label for root window
label1 = Label(root, text = "This is the root window")

# define a function for 2nd toplevel
# window which is not associated with
# any parent window
```

```
def open_Toplevel2():
    # Create widget
    top2 = Toplevel()

    # define title for window
    top2.title("Toplevel2")

    # specify size
    top2.geometry("200x100")

    # Create label
    label = Label(top2, text = "This is a Toplevel2
    window")

    # Create exit button.
    button = Button(top2, text = "Exit", command = top2.
    destroy)
    label.pack()
    button.pack()

    # Display until closed manually.
    top2.mainloop()

# define a function for 1st toplevel
# which is associated with root window.
def open_Toplevel1():

    # Create widget
    top1 = Toplevel(root)

    # Define title for window
    top1.title("Toplevel1")

    # specify size
    top1.geometry("200x200")

    # Create label
    label = Label(top1, text = "This is a Toplevel1
    window")

    # Create Exit button
    button1 = Button(top1, text = "Exit", command = top1.
    destroy)

    # create button to open toplevel2
    button2 = Button(top1, text = "open toplevel2",
                command = open_Toplevel2)

    label.pack()
    button2.pack()
```

```
    button1.pack()

    # Display until closed manually
    top1.mainloop()
# Create button to open toplevel1
button = Button(root, text = "open toplevel1", command =
open_Toplevel1)
label1.pack()

# position the button
button.place(x = 155, y = 50)

# Display until closed manually
root.mainloop()
```

12.4.2 Frames

Box modules are other widget cabinets. While you can connect mouse and keyboard events to callback devices, frames have restricted choices and no other methods than normal widget choices [102]. A master for a set of widgets is one of the more widely used with a frame that is operated by a geometry operator. Figure 12.6 illustrates this. For each round of the test the second example of a frame is seen in Figure 12.7.

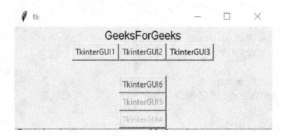

FIGURE 12.6
Frame widget.

```
from tkinter import *
root = Tk()
root.geometry("300x150")

w = Label(root, text = 'hello GUI', font = "50")
w.pack()
```

```
frame = Frame(root)
frame.pack()

bottomframe = Frame(root)
bottomframe.pack(side = BOTTOM)
b1_button = Button(frame, text="TkinterGUI1", fg = "red")
b1_button.pack(side = LEFT)

b2_button = Button(frame, text="TkinterGUI2", fg = "brown")
b2_button.pack(side = LEFT)

b3_button = Button(frame, text="TkinterGUI3", fg = "blue")
b3_button.pack(side = LEFT)
b4_button = Button(frame, text="TkinterGUI4", fg = "pink")
b4_button.pack(side = BOTTOM)

b5_button = Button(frame, text="TkinterGUI5", fg = "orange")
b5_button.pack(side = BOTTOM)

b6_button = Button(frame, text="TkinterGUI6", fg = "green")
b6_button.pack(side = BOTTOM)

root.mainloop()
```

The frame appearance may be changed, similarly to buttons and labels, by picking a relief form and adding the desired boundary length. In reality, the distinction between these widgets may be difficult to see (see Figure 12.7). For this purpose it is advisable to reserve separate decoration for individual widgets and not to allow for example the decorations of a sticker for a button to be used.

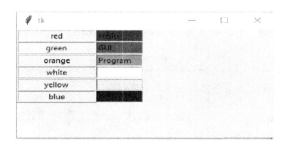

FIGURE 12.7
Advance frame widget.

```
import tkinter as tk

colours = ['red', 'green', 'orange', 'white', 'yellow',
'blue']
r = 0
for c in colours:
    tk.Label(text=c, relief=tk.RIDGE, width=15).
    grid(row=r, column=0)
    tk.Entry (bg=c, relief=tk.SUNKEN, width=10).
    grid(row=r, column=1)
    r = r + 1

tk.mainloop ()
```

12.4.3 Labels

Mark buttons for the text or picture view are included. Labels can provide multi-line text, but only a single font can be used [103]. You can divide a text string that fits the remaining space, or insert line feeds in a power break string. Figure 12.8 displays many stickers.

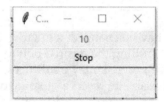

FIGURE 12.8
Label widget.

Although marks are not meant for use in user interaction, mouse and keyboard events may be linked to callbacks. For some applications, this can be used as a "cheap" press.

```
import tkinter as tk

counter = 0
def counter_label(label):
  def count():
    global counter
    counter +=1
    label.config(text=str(counter))
```

```
      label.after(1000, count)
   count()

root = tk.Tk()
root.title("Counting Seconds")
label = tk.Label(root, fg="green")
label.pack()
counter_label(label)
button = tk.Button(root, text='Stop', width=25,
command=root.destroy)
button.pack()
root.mainloop()
```

12.4.4 Buttons

Strictly speaking, the keys are cursor responding icons. When your button is activated, you connect to a method call or callback [104]. To prevent users from turning on a button, icons may be deactivated. The button widget can include text or images (which could occupy several lines). Buttons should be in the button category, so you can use the key to access them. Figure 12.9 displays the basic keys.

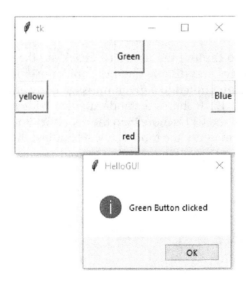

FIGURE 12.9
Button widgets.

```
import tkinter
from tkinter import *
from tkinter import messagebox

top = Tk()
top.geometry("300x150")
def click():
    messagebox.showinfo("HelloGUI", "Green Button clicked")
a = Button(top, text="yellow", activeforeground="yellow",
activebackground="orange", pady=10)
b = Button(top, text="Blue", activeforeground="blue",
activebackground="orange", pady=10)
# adding click function to the below button
c = Button(top, text="Green", command=click,
activeforeground = "green", activebackground="orange",
pady=10)
d = Button(top, text="red", activeforeground="yellow",
activebackground="orange", pady=10)

a.pack(side = LEFT)
b.pack(side = RIGHT)
c.pack(side = TOP)
d.pack(side = BOTTOM)
top.mainloop()
```

12.4.5 Entry

The key widgets to capture the user's feedback are the entry widgets. You can also send images and disable them to avoid a user modifying its value. Entering widgets are limited to a given message body and are only possible in one font. Figure 12.10 shows a standard entry widget [105]. The widget scrolls the output if the text is more than the necessary counter space inserted in the widget. You may use the mouse wheel to adjust the apparent location. You may also connect scrolling measures to your mouse or device using the scrolling method of the widget.

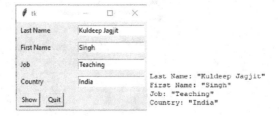

FIGURE 12.10
Entry widget.

```
import tkinter as tk

fields = 'Last Name', 'First Name', 'Job', 'Country'

def fetch(entries):
    for entry in entries:
        field = entry[0]
        text = entry[1].get()
        print('%s:"%s"' %(field, text))

def makeform(root, fields):
    entries = []
    for field in fields:
        row = tk.Frame(root)
        lab = tk.Label(row, width=15, text=field,
        anchor='w')
        ent = tk.Entry(row)
        row.pack(side=tk.TOP, fill=tk.X, padx=5, pady=5)
        lab.pack(side=tk.LEFT)
        ent.pack(side=tk.RIGHT, expand=tk.YES, fill=tk.X)
        entries.append((field, ent))
    return entries

if __name__ == '__main__':
    root = tk.Tk()
    ents = makeform(root,fields)
    root.bind('<Return>', (lambda event, e=ents:
    fetch(e)))
    b1 = tk.Button(root, text='Show', command=(lambda
    e=ents: fetch(e)))
    b1.pack(side=tk.LEFT, padx=5,pady=5)
    b2 = tk.Button(root, text='Quit', command=root.quit)
    b2.pack(side=tk.LEFT, padx=5, pady=5)
    root.mainloop()
```

12.4.6 Radio Buttons

The radio button widget will soon have to be replaced! Car radios with electronic keys are becoming rare and difficult to use to explain the GUI [106]. However, the concept is that all options are exclusive to delete all already chosen buttons by clicking one button. Radio buttons can view text or images in a similar way to button widgets and they have text spanning several lines, but only in one font. A typical radio button can be seen in Figure 12.11. In the community, you usually allocate a single attribute to all the radio keys.

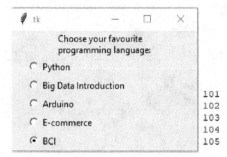

FIGURE 12.11
Radio button widget.

```python
import tkinter as tk

root = tk.Tk()

v = tk.IntVar()
v.set(1)

languages = [("python", 101),
             ("Big Data Introduction", 102),
             ("Arduino", 103),
             ("E-commerce", 104),
             ("BCI", 105)]

def ShowChoice():
    print(v.get())

tk.Label(root, text="""Choose your favourite
programming language:""",
        justify = tk.LEFT,
        padx = 20).pack()

for language, val in languages:
    tk.Radiobutton(root,
        text=language,
        padx = 20,
        variable = v,
        command = ShowChoice,
        value = val).pack(anchor = tk.W)
    root.mainloop()
```

12.4.7 Check Buttons

For choosing one or more objects, check button widgets are used. There is no connection between check keys, unlike radio keys. For either texts or photos you can load the search buttons [105]. Check buttons should normally include an `Int Var` variable which allows you to define the check button's status, which is allocated to a variable option. You may also (or instead of) bind a callback to the button that is called when you press the button. The design of check buttons on UNIX and Windows is very different; usually, Unix uses a fill color to mark the range, while Windows uses a checkmark. Figure 12.12 indicates the appearance in Windows.

FIGURE 12.12
Check button widgets.

```
from tkinter import *
class Checkbar(Frame):
    def __init__(self, parent=None, picks=[], side=LEFT,
    anchor=W):
        Frame.__init__(self,parent)
        self.vars = []
        for pick in picks:
            var = IntVar()
            chk = Checkbutton(self, text=pick,
            variable=var)
            chk.pack(side=side, anchor=anchor, expand=YES)
            self.vars.append(var)
    def state(self):
        return map((lambda var: var.get()), self.vars)

if __name__=='__main__':
    root = Tk()
    lng = Checkbar(root, ['python', 'BigData', 'BCI',
        'Arduino'])
    tgl = Checkbar(root, ['English', 'German'])
    lng.pack(side=TOP, fill=X)
    tgl.pack(side=LEFT)
    lng.config(relief=GROOVE, bd=2)

    def allstates():
```

```
      print(list(lng.state()), list(tg1.stste()))
Button(root, text='Quit', command=root.quit).
pack(side=RIGHT)
Button(root, text='Peek', command=allstates).
pack(side=RIGHT)
root.mainloop()
```

12.4.8 Messages

The message overlay offers a simple way to view multi-line text. For the full post, you should use one font and one color in the foreground/background (see Figure 12.13). The widget has regular GUI methods, which provide an example of this GUI [97].

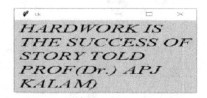

FIGURE 12.13
Message widget.

```
import tkinter as tk
master = tk.Tk()
whatever_you_do = "HARDWORK IS THE SUCCESS OF STORY TOLD
PROF(Dr.) apj KALAM"
msg = tk.Message(master, text = whatever_you_do)
msg.config(bg='lightgreen', font=('times', 24, 'italic'))
msg.pack()
tk.mainloop()
```

12.4.9 List Boxes

List box widgets display a list of consumer selectable properties and the user is allowed to pick a single object in the list by default [100]. Figure 12.14 shows a basic case. Use the selection mode feature for the widget to allow different objects and other assets to include additional binds. For clarification on adding scrolling power to the list window, see "Scroll Bar" shown in Figure 12.4.

FIGURE 12.14
List box widgets.

```
from tkinter import *
top = Tk()
listbox = Listbox(top, height = 10,
        width = 15,
        bg = "grey",
        activestyle = 'dotbox',
        font = "Helvetica",
        fg = "blue")

top.geometry("300x250")
label = Label(top, text = "books")

listbox.insert(1, "Python")
listbox.insert(2, "BCI")
listbox.insert(3, "Arduino")
listbox.insert(4, "Bigdata")
listbox.insert(5, "E-commerce")
label.pack()
listbox.pack()
top.mainloop()
```

12.5 Canvases

Various buttons are canvases. You may use them not only for drawing abstract shapes using circles, ovals, polygons and rectangles but also for extremely specific positioning of pictures and bit-maps on the canvas [101]. Also, all widgets (e.g. keys and list boxes) can be put inside a canvas and have a mouse or keypad attached to them. You can see several cases, where canvas widgets are used to provide a free-flowing container for several purposes.

Figure 12.15 explains most of the open services. One factor that can support or hinder canvas widgets is that objects have been painted on every object on the canvas. If required, you may later adjust the order of canvas objects.

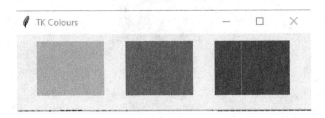

FIGURE 12.15
Canvas widget.

```
from tkinter import Tk, Canvas, Frame, BOTH
class Example(Frame):
    def __init__(self):
        super().__init__()
        self.initUI()
    def initUI(self):
        self.master.title("TK Colours")
        self.pack(fill=BOTH, expand=1)
        canvas = Canvas(self)
        canvas.create_rectangle(30, 10, 120, 80,
            outline="#fb0", fill="#fb0")
        canvas.create_rectangle(150, 10, 240, 80,
            outline="#f50", fill="#f50")
        canvas.create_rectangle(270, 10, 370, 80,
            outline="#05f", fill="#05f")
        canvas.pack(fill=BOTH, expand=1)
def main():

    root = Tk()
    ex = Example()
    root.geometry("400x100+300+300")
    root.mainloop()
if __name__=='__main__':
    main()
```

12.6 Geometry Managers

Geometry administration is a difficult topic, and there is a great deal of communication between widget containers, windows and window managers. The objective is to construct one or several control devices (some programmers choose children's widgets and relatives) as subordinates to a master widget [103]. Usually, master widgets are containers such as frames or bathrooms, but most widgets can be masters. For example, we place a button at the end of an image. We wish to track the widget's behavior by including more widgets or by growing or extending the window, as well as identifying slaves inside masters. A slave widget demands width and height sufficient for showing its contents, which starts the negotiating process. This is based on many variables. For example, a button calculates the appropriate size from the size and weight of a text showing as a mark. First, the professional widget calculates the space available along with its geometry manager to satisfy the necessary slave specifications. The required space can be more or less than the necessary space, allowing the widgets to pinch, extend or overlap based on the geometry manager. The room inside a master must be distributed between all peer containers following the configuration of the frame. The results depend on the peer widget's geometry manager.

Finally, we discuss the top button (usually the top shell) and the window manager. The appropriate measurements are used for determining the final scale and position in which the widgets should be drawn after negotiations. In certain instances, there may be inadequate capacity for all widgets to be shown. When a window is enabled, and when these discussions have ended, it continues again when either of the controls adjusts the setting (such as if the text on the button changes) or the user resizes the window. Luckily, using geometry managers is much better than learning about them! When a computer is designed, a variety of different programs may be implemented. The packer and to a smaller degree the grid have the opportunity to assess a window's final size using the geometry manager. This is valuable when a window is dynamically generated and the population of widgets is hard to anticipate.

12.7 Loan Calculators

The window adjusts size by inserting or deleting widgets from the view through this method. Alternately, the builder may use a location on a set frame [99]. It depends on the desired result. Let's begin by looking at the regular manager.

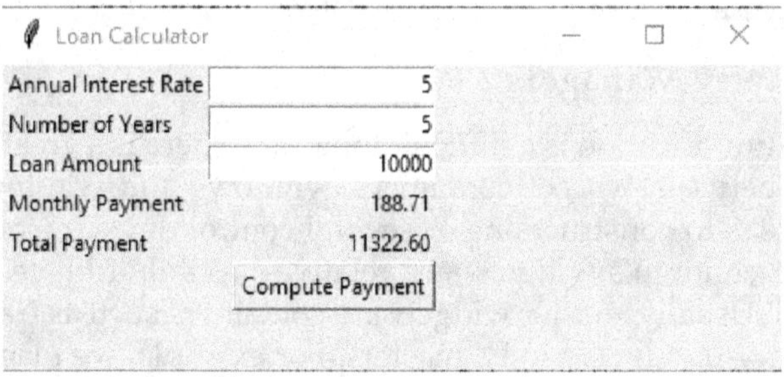

FIGURE 12.16
Loan calculation interface.

To create a tkinter:

1. Load the tkinter module;
2. Build the key (container) window;
3. Add some widgets to the main window;
4. On the dashboard, add the event key.

Let's see how to use the Python GUI Tkinter library to build a loan calculator, which will calculate the overall and recurring payments based on the
amount of the loan's interest rate.

```python
# Import tkinter
from tkinter import *
class LoanCalculator:
    def __init__(self):
        window = Tk() # Create a window
        window.title("Loan Calculator") # Set title
        # create the input boxes.
        Label(window, text = "Annual Interest Rate").
        grid(row = 1,        column = 1, sticky = W)
        Label(window, text = "Number of Years").grid(row
        = 2,                 column = 1, sticky = W)
        Label(window, text = "Loan Amount").grid(row = 3,
        column = 1, sticky = W)
        Label(window, text = "Monthly Payment").grid(row
        = 4,                 column = 1, sticky = W)
        Label(window, text = "Total Payment").grid(row =
            5,               column = 1, sticky = W)

        # for taking inputs
        self.annualInterestRateVar = StringVar()
```

```
            Entry(window, textvariable = self.
            annualInterestRateVar, justify = RIGHT).grid(row
            = 1, column = 2)
            self.numberOfYearsVar = StringVar()
            Entry(window, textvariable = self.
            numberOfYearsVar, justify = RIGHT).grid(row = 2,
            column = 2)
            self.loanAmountVar = StringVar()
            Entry(window, textvariable = self.loanAmountVar,
            justify = RIGHT).grid(row = 3, column = 2)
            self.monthlyPaymentVar = StringVar()
            lblMonthlyPayment = Label(window, textvariable =
            self.monthlyPaymentVar).grid(row = 4,
            column = 2, sticky = E)

            self.totalPaymentVar = StringVar()
            lblTotalPayment = Label(window, textvariable =
            self.totalPaymentVar).grid(row = 5,
            column = 2, sticky = E)

            # create the button
            btComputePayment = Button(window, text = "Compute
            Payment",      command = self.computePayment).grid(
                            row = 6, column = 2, sticky = E)
            window.mainloop() # Create an event loop

    # compute the total payment.
    def computePayment(self):

            monthlyPayment = self.getMonthlyPayment(
            float(self.loanAmountVar.get()),
            float(self.annualInterestRateVar.get()) / 1200,
            int(self.numberOfYearsVar.get()))

            self.monthlyPaymentVar.set(format(monthlyPayment,
            '10.2f'))
            totalPayment = float(self.monthlyPaymentVar.
            get()) * 12 \
                        * int(self.numberOfYearsVar.
                        get())

            self.totalPaymentVar.set(format(totalPayment,
            '10.2f'))

    def getMonthlyPayment(self, loanAmount,
    monthlyInterestRate, numberOfYears): # compute the
    monthly payment.
```

```
        monthlyPayment = loanAmount * monthlyInterestRate
        / (1- 1 / (1 + monthlyInterestRate) **
        (numberOfYears * 12))
        return monthlyPayment;
        root = Tk() # create the widget
   # call the class to run the program.
LoanCalculator()
```

Code explanation:

- The toolkit of Tkinter is included. We import the entire Tkinter package in the first line here in this case. Next, we build a class called LoanCalculator which contains its data member and membership functionality.

- def init (self) in a Python class is a special process, a class-building system. We then develop a window using Tk. The marking feature produces an input display box and uses a grid solution to generate a layout, like a line.

12.8 Displaying Images

You may use the show function on a picture object to view it or to view a picture in Python Pillow [98].

The display protocol copies the image to a temporary file and then unlocks the automatic image display program. The temporary file is erased until the execution of the program is finished.

Example: Pillow shows or monitors the picture in the example below, and we can send a picture to the user using a display method as in Figure 12.17.

```
from PIL import Image

#read the image
im = Image.open("sample-image.png")

#show image
im.show()
```

FIGURE 12.17
Display image.

We are using Windows PC in this scenario; Photos is the normal framework for opening Bitmap Image (BMP) files. Therefore, the show pillow method shows a picture using the software images.

12.9 Menus

A common way to encourage the user to select the activities of an application is by menu widgets. Menus can be very noisy to create, particularly if the cascades go out on different levels (the most popular approach is to try to develop menus so that you don't have to go out on all three different levels to reach any functionality) [103]. For menu design, Tkinter offers versatility to various fonts, icons and bitmaps, and search buttons and radio keys. A menu can be generated in many different schemes. The example seen in Figure 12.18 is a means of constructing a menu; an alternative scheme to construct the same menu online as altmenu.py is possible.

```
from tkinter import *
from tkinter.filedialog import askopenfilename

def NewFile():
    print("New File!")
def OpenFile():
    name = askopenfilename()
    print(name)
```

```
def About():
    print("This is a simple example of a menu")

root = Tk()
menu = Menu(root)
root.config(menu=menu)
filemenu = Menu(menu)
menu.add_cascade(label="file", menu=filemenu)
filemenu.add_command(label="New", command=NewFile)
filemenu.add_command(label="Open", command=OpenFile)
filemenu.add_separator()
filemenu.add_command(label="Exit", command=root.quit)

helpmenu = Menu(menu)
menu.add_cascade(label="Help", menu=helpmenu)
helpmenu.add_command(label="About...", command=About)

mainloop()
```

FIGURE 12.18
Menu display.

12.10 Popup Menus

A popup screen is like a normal menu, except it has no menu bar and can be rotated on the display. It's like a menu that appears [105].

In Figure 12.19, the same phenomenon produces a popup screen. First, a menu instance can be created, and then items can be inserted. You can eventually add a button to the menu for a case.

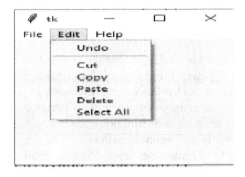

FIGURE 12.19
Popup menu.

```
from tkinter import Toplevel, Button, Tk, Menu

top = Tk()
menubar = Menu(top)
file = Menu(menubar, tearoff=0)
file.add_command(label="New")
file.add_command(label="Open")
file.add_command(label="Save")
file.add_command(label="Save as...")
file.add_command(label="Close")
file.add_separator()
file.add_command(label="Exit", command=top.quit)
menubar.add_cascade(label="File", menu=file)

edit = Menu(menubar, tearoff=0)
edit.add_command(label="Undo")
edit.add_separator()
edit.add_command(label="Cut")
edit.add_command(label="Copy")
edit.add_command(label="Paste")
edit.add_command(label="Delete")
edit.add_command(label="Select All")
menubar.add_cascade(label="Edit", menu=edit)
help = Menu(menubar, tearoff=0)
help.add_command(label="About")
menubar.add_cascade(label="Help", menu=help)

top.config(menu=menubar)
top.mainloop()
```

12.11 The Mouse, Key Events and Bindings

You can create complicated GUI programs, regardless of whether the program operates in a Unix, Windows or Macintosh environment without understanding much about the basic event mechanism [101]. However, if you know how to order and manage events in your program, it is typically easier to create an implementation that functions the way you want. Readers who are familiar with events and event handlers in X or windows messages may want to look ahead to "Tkinter incidents," as that detail is Tkinter unique.

What are Events?

Events provide alerts (in Windows language messages) which are sent to the client code by the windowing mechanism (e.g. X-server for X). You claim that something happened or altered the status of any managed object, either because of the user entry or because the code demanded a modification to the server [97]. Applications usually do not immediately accept incidents. You will not however be aware of the activities demanded implicitly by your applications or requests created through widgets. For example, when a button is pressed, you can assign a callback command; the widget links an occurrence to the callback. An occurrence that is typically treated somewhere may also be alerted. It may be a positive thing, but the nature of complex number devices can also be impaired, so it has to be done with caution. The program may alter its behavior, usually through monitors and windows. In any case it is in the queue of an operation. Events are usually excluded from the main loop by a feature name. You'll usually use the Tkinter Haul Loop, but if you have particular requirements you can supply a special Haul Loop (such as a threaded program that may control internal locks in a way that prohibits the normal scheme from being used). Tkinter allows you unimplemented access to events so you don't need to know anything about the underlying event handler, for example, in the following short clip, to detect when the cursor reaches a frame in Figure 12.20.

```
import turtle

turtle.setup(400,500)
wn = turtle.Screen()
wn.title("Green traffic light")
wn.bgcolor("lightgreen")
tess = turtle.Turtle()
```

```
def draw_housing():
    tess.pensize(3)
    tess.color("black","darkgrey")
    tess.begin_fill()
    tess.forward(80)
    tess.left(90)
    tess.forward(200)
    tess.circle(40, 180)
    tess.forward(200)
    tess.left(90)
    tess.end_fill()

draw_housing()
tess.penup()
tess.forward(40)
tess.left(90)
tess.forward(50)
tess.shape("circle")
tess.shapesize(3)
tess.fillcolor("green")

state_num = 0

def advance_state_machine():
    global state_num
    if state_num == 0:
        tess.forward(70)
        tess.fillcolor("orange")
        state_num = 1
    elif state_num == 1:
        tess.forward(70)
        tess.fillcolor("red")
        state_num = 2
    else:
        tess.back(140)
        tess.fillcolor("green")
        state_num = 0

wn.onkey(advance_state_machine, "space")

wn.listen()
wn.mainloop()
```

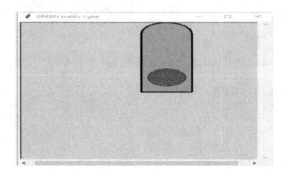

FIGURE 12.20
Traffic light.

The frame binding mechanism is used to bind the callback data in some cases. The message is written once the mouse crosses the boundary from the outside to the inside.

Event propagation

Events are connected to a frame, typically defined as the event's source window. If no client has reported the source window of a specific case, the event will spread across the window structure before a client has either found the window that is entered by a client, or discovers a window preventing the spread of an event, or reaches the root window. The occurrence is missed until it enters the root frame. There is just a proliferation of system events arising from a switch, pointer movement and mouse clicks. Additional events, including display and setup events, must be specifically documented.

Event types

According to X, which represents events, case masks, events are divided into many groups. When operating on a Windows system, Tk maps Windows events to the same masks. Table 12.1 displays the case masks found by Tk (and thus Tkinter).

- **Keyboard events**. When you push a key, you create a keypress event; when the key is released, you can trigger a key release event. Keyboard activities are created by moving buttons, such as SHIFT and Power.

- **Pointer events**. When mouse buttons are pressed or the cursor is pushed, events are created by pressing the button, releasing it and notifying the cursor. The event window is the lowest hierarchy window unless there is a pointer log, in which case the window to

TABLE 12.1

Types of Events

NoEventMask	StructureNotifyMask	Button3MotionMask
KeyReleaseMask	SubstructureNotifyMask	Button5MotionMask
ButtonReleaseMask	FocusChangeMask	KeymapStateMAsk
LeaveWindowMask	ColormapChangeMask	VisibilityChangeMask
PonterMotionHintMAsk	KeyPressMask	ResizeRedirectMask
Button2MotionMask	ButtonPressMask	SubstructureRedirectMAsk
Button4MotionMask	EnterWindowMask	PropertyChangeMask
ButtonMotionMask	PointerMotionMask	OwnerGrabButtonMask
ExposureMask	Button1MotionMask	

the data is listed. Editing keys can be applied to pointer events like keyboards.

- **Crossing events**. An enter notify or leave notify event is created whenever the pointer enters or leaves a window cap. Whether the crossover was due to the pointer shifting or due to the change in the stacking order of the windows does not matter. For example if behind another window a window containing the pointer is lowered and the pointer is now in the upper

- **Focus events**. Described as the target window, it is the window that acquires keyboard events. Whenever the focus window varies, focus in and focus out events are created. Management of focus events is somewhat trickier than controlling pointer events because the arrow does not have to be in the space where the focus events are obtained. You typically don't have to manage focal events yourself, since by pressing the TAB key, you can switch attention between widgets.

- **Exposure events**. An exposure event is produced anytime that a window or part of a window is visible. Usually, you cannot plan exposure events in Tkinter GUIs, except if you have a very unique drawing to help these activities. Events settings configure alert events which are created when the height, location or boundary of a window changes. When the stacking order of the windows varies, a configure alert event is generated. Gravity, map/unmapped, reflect and illumination are other kinds of configuration cases.

- **Color map events**. A color map alert event is created when a new color map is enabled. This can be used to stop distracting color map flickering that can arise if a color map is activated by another program. Most apps, however, do not explicitly monitor their color charts.

12.12 Animations

Tkinter is one of Python's most popular GUI toolkits. It is available in regular installation. This kit can then be used without any extra modules being installed. You can use Tkinter for building full-user interfaces or building basic desktop players with a versatile GUI library [96].

This segment includes one hand-by-one animation tutorial that uses the program Python Tkinter. To draw our animations, we use the canvas of the Tkinter kit. Note that Python 3.6 or higher is required for the following application.

A basic animation displaying the motion of a ball around the screen is created using the following Python software and is shown in Figure 12.21.

The steps in the procedure are:

- Create and view program's main window.
- Create and mount to main window canvas; we will draw a circle (our ball) within the canvas.
- Draw and animate.

Let us begin with the importing of the time and Tkinter modules. In order to postpone drawing of the animation screens, we can use the time module

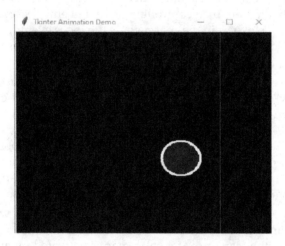

FIGURE 12.21
Animation demo.

```
        import tkinter
import time

# width of the animation window
animation window_width=800
# height of the animation window
animation window height=600
# initial x position of the ball
animation_ball_start_xpos = 50
# initial y position of the ball
animation_ball_start_ypos = 50
# radius of the ball
    animation_ball_radius = 30
# the pixel movement of ball for each iteration
animation_ball_min_movement = 5
# delay between successive frames in seconds animation_
refresh_secanda = 0.01

# The main window of the animation
def create_animation_window():
   window = tkinter.Tk()
   window.title("Tkinter Animation Demo")
   # Uses python 3.6+ string interpolation
   window.geometry(f'{animation_window_width}
   x{animation_window_height}')
   return window

# Create a canvas for animation and add it to main window
def create_animation_canvas(window):
   canvas = tkinter.Canvas(window)
   canvas.configure (bg="black")
   canvas.pack(fill="both", expand=True)
   return canvas

# Create and animate ball in an infinite loop
def animate_ball(window, canvas,xinc, yinc):
   ball = canvas.create_oval(animation_ball_
   start_xpos-animation_ball_radius, animation_
   ball_start_ypos-animation_ball_radius,
   animation_ball_start_xpos+animation_ball_radius,
   animation_ball_start_ypos+animation_ball_radius,
   fill="blue", outline="white", width=4)
   while True:
     canvas.move (ball, xinc, yinc)
     window.update()
     time.sleep(animation_refresh_seconds)
```

```
        ball_pos = canvas.coords(ball)
        # unpack array to variables
        x1,y1,xr,yr = ball_pos
        if x1 < abs(xinc) or xr >
        animation_window_width-abs(xinc):
            xinc = -xinc
        if y1 < abs(yinc) or yr >
        animation_window_height-abs(yinc):
            yinc = -yinc
    # The actual execution starts here
animation_window = create_animation_window()
animation_canvas = create_animation_canvas(animation_
    window)
animate_ball(animation_window, animation_canvas,
    animation_ball_min_movement, animation_ball_min_
    movement)
```

12.13 Scrollbars

Each widget enabling scrolling like text, map and list box widgets can be used with a scrollbar widget [97]. Trying to associate a scrollbar widget with an external widget is as simple as adding calls to each widget to display them. Naturally, you do not have to co-locate them, but if you do not, you may end up with odd GUIs! The standard implementation is shown in Figure 12.22.

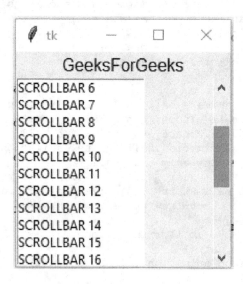

FIGURE 12.22
Scrollbar widget.

```
from tkinter import *
root = Tk()
root.geometry("150x200")
w = Label(root, text = 'GeeksForGeeks', font = "50")
w.pack()
scroll_bar = Scrollbar(root)
scroll_bar.pack(side = RIGHT, fill = Y)
mylist = Listbox(root, yscrollcommand = scroll_bar.set)
for line in range(1, 26):
    mylist.insert(END, "SCROLLBAR" + str(line))
    mylist.pack(side = LEFT, fill = BOTH)
    scroll_bar.config(command = mylist.yview)
    root.mainloop()
```

12.14 Standard Dialog Boxes

Dialogs

Dialogs are only special types of cases. Generally speaking, dialogs send the user alerts or error messages, ask questions or accumulate a small set of user values (typically one value). You might say both types are dialogs. However, they are typically a good model to replace the other unidentified ones [98]. The modalities may be app-wide or system-wide, so you must make sure that system-modal dialogs are reserved for conditions that the user must understand before any further contact is feasible.

Tkinter includes a dialog pack. But the downside is that it uses X bitmaps in the case of bugs, alerts and other icons. The module tkSimpleDialog determines the string request, the integer request and the float request to gather strings, int and floats. The module tkMessageBox specifies convenient functions like display information, display alert and show error. The architectural style-specific icons for tkMessageBox are used so that the icons look perfect on all viewing platforms.

Standard dialogs

These are easy to use template dialogues. Various tools, including show error, show warning and requestretrycancel, are available in tkMessageBox. This example shows the use of only one kind of dialog available (on demand):

```
import tkinter as tk
from tkinter.colorchooser import askcolor

def callback():
    result = askcolor(color="#6A9662",title = "Bernd's
    Colour Chooser")
    print (result)

root = tk.Tk()
tk.Button(root, text='Choose Color', fg = "darkgreen",
command=callback).pack(side=tk.LEFT, padx=10)
tk.Button(text='Quit',command=root.quit, fg="red").
pack(side=tk.LEFT, padx=10)
tk.mainloop()
```

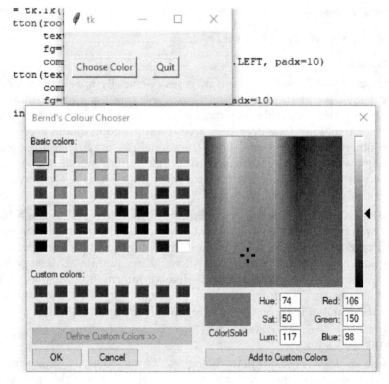

FIGURE 12.23
Standard dialog box.

Code comments

The title and prompt (because it is a query dialog) are given in the first two arguments in Figure 12.23. The first sets the default behavior (the action connected with pressing the RETURN key) to the button with the selected number. For example: OK for the O.K. button, cancel for the CANCEL button; the regular dialogues return the pressed button as a string.

12.15 Conclusion

This chapter has shown us what we can know about the simple GUI functionality of our Hello World program. By using Tkinter's built-in widgets, we learned how to build a window that includes different GUI components. Tkinter is excellent for small, quick GUI applications, and since it runs on more platforms than any other Python GUI toolkit, it is a good choice where portability is the prime concern. This chapter started by creating a basic layout window using Tkinter. After that, we explored the different widgets that there are in the Tkinter library and used them to create a form, as well as how different widgets on the application window using the function can be created.

13

Python Exception Handling: GUI Programming Using Tkinter

13.1 Introduction

Sometimes a Python program does not execute at all or executes so far then generates unexpected output or behaves abnormally. These situations occur when there are syntax errors, runtime errors or logical errors in the code. In Python, exceptions are errors thatget triggered automatically. However, exceptions can be forcefully triggered and handled through program code. In this chapter, we will learn about exception handling in Python programs [107].

13.2 Exception Examples

Before we get into why exception handling is essential, and the types of built-in exceptions that Python supports, it is necessary to understand that there is a subtle difference between an error and an exception.

Errors cannot be handled, while Python exceptions can be handled at run time. An error can be a syntax (parsing) error, while there can be many types of exceptions that occur during execution and are not unconditionally inoperable. An error might indicate critical problems that a reasonable application should not try to catch, while an exception might indicate conditions that an application should try to catch. Errors are a form of unchecked exception and are irrecoverable, such as an OutOfMemoryError, which a programmer should not try to handle.

Exception handling makes your code more robust and helps prevent potential failures that would cause your program to stop in an uncontrolled manner. Imagine you have written code which is deployed in production but terminates due to an exception. Your client would not appreciate that, so it's better to handle the particular exception beforehand and avoid the chaos.

Errors can be of various types.

Syntax errors: Syntax errors are often called parsing errors, and are predominantly caused when the parser detects a syntactic issue in your code.

Let's take an example to understand it.

```
a = 8
b = 10
c = a b
  File "<ipython-input-8-3b3ffcedf995>", line 3
    c = a b
        ^
SyntaxError: invalid syntax
```

The above arrow indicates when the parser ran into an error while executing the code. The token preceding the arrow causes the failure. To rectify such fundamental errors, Python will do most of the job for you since it will print the file name and the line number at which the error occurred.

Out of Memory Error: Memory errors are mostly dependent on your systems RAM and are related to a Heap. If you have large objects (or) referenced objects in memory, then you will see the OutofMemoryError. This can be due to various reasons:

- Using a 32-bit Python architecture (the maximum memory allocation given is very low, between 2GB and 4GB).
- Loading a very large data file.
- Running a machine learning/deep learning model, or many others.

You can handle the memory error with the help of exception handling, which is a fallback exception for when the interpreter entirely runs out of memory and must immediately stop execution. In these rare instances, Python raises an OutofMemoryError, allowing the script to somehow catch itself and break out of the memory error and recover itself [108].

However, since Python adopts to the memory management architecture of the C language (malloc() function), it is not certain that all processes of the script will recover – in some cases, a MemoryError will result in an unrecoverable crash. Hence, it is not good practice to use exception handling for such an error.

Recursion error: This is related to the stack and occurs when you call functions. As the name suggests, a recursion error transpires when too many methods, one inside another, are executed (one with an infinite recursion), which becomes limited by the size of the stack.

All your local variables and methods that call associated data will be placed on the stack. For each method call, one stack frame will be created, and local as well as method call relevant data will be placed inside that stack frame. Once the method execution is completed, the stack frame will be removed.

To reproduce this error, let's define a function that will be recursive, meaning it will keep calling itself as an infinite loop method call. You will see a StackOverflow or a Recursion Error because the stack frame will be populated with method data for every call, but it will not be freed.

```
def recursion():
  return recursion()
recursion()
```

```
RecursionError                 Traceback (most recent call last)

<ipython-input-3-c6e0f7eb0cde> in <module>
----> 1 recursion()

<ipython-input-2-5395140f7f05> in recursion()
      1 def recursion():
----> 2      return recursion()

... last 1 frame repeated, from the frame below ...
<ipython-input-2-5395140f7f05> in recursion()
      1 def recursion():
----> 2      return recursion()

RecursionError: maximum recursion depth exceeded
```

Indentation error. This is similar in spirit to the syntax error and falls under it. However, it is specific only to the indentation related issues in the script.

So let's use a quick example to understand an indentation error.

```
for i in range(10):
print('Hello world')
  File "<ipython-input-6-628f419d2da8>", line 2
  print('Hello world')
     ^
IndentationError: expected an indented block
```

Exceptions. Even if the syntax of a statement or expression is correct, it may still cause an error when executed. Python exceptions are errors that are detected during execution and are not unconditionally fatal: you will soon learn how to handle them in Python programs. An exception object is created when a Python script raises an exception. If the script explicitly doesn't handle the exception, the program will be forced to terminate abruptly [109].

Programs usually do not handle exceptions, and result in error messages as shown here:

```
Type Error
a = 2
b = 'DataCamp'
a + b
----------------------------------------------------------------

TypeError                         Traceback (most recent call last)
<ipython-input-7-86a706a0ffdf> in <module>
      1 a = 2
      2 b = 'DataCamp'
----> 3 a + b

TypeError: unsupported operand type(s) for +: 'int' and 'str'
Zero Division Error
100 / 0
----------------------------------------------------------------

ZeroDivisionError                         Traceback (most
recent call last)
<ipython-input-43-e9e866a10e2a> in <module>
----> 1 100 / 0

ZeroDivisionError: division by zero
```

There are various types of Python exceptions, and the type is printed as part of the message: in the above two examples they are ZeroDivisionError and TypeError. Both the error strings printed, as the exception type, are the name of Python's built-in exception.

The remaining part of the error line provides the details of what caused the error, based on the type of exception.

Before you start learning the built-in exceptions, let's just quickly revise the four main components of exception handling:

- **Try**: This will run the code block in which you expect an error to occur.
- **Except**: Here, you will define the type of exception you expect in the try block (built-in or custom).
- **Else**: If there isn't any exception, then this block of code will be executed (consider this as a remedy or a fallback option if you expect a part of your script to produce an exception).
- **Finally**: Irrespective of whether there is an exception or not, this block of code will always be executed.

You will learn about the common types of exceptions and also learn to handle them with the help of exception handling.

Keyboard interrupt error. The KeyboardInterrupt exception is raised when you try to stop a running program by pressing ctrl+c or ctrl+z in a command line or interrupting the kernel in Jupyter Notebook. Sometimes you might not intend to interrupt a program, but by mistake it happens, in which case using exception handling to avoid such issues can be helpful.

In the below example, if you run the cell and interrupt the kernel, the program will raise a KeyboardInterrupt exception inp = input (). Let's now handle the exception.

```
try:
    inp = input()
    print ('Press Ctrl+C or Interrupt the Kernel:')
except KeyboardInterrupt:
    print ('Caught KeyboardInterrupt')
else:
    print ('No exception occurred')
Caught KeyboardInterrupt
Standard Error
```

Let's learn about some of the standard errors that can often occur while programming.

Arithmetic Error:

- Zero Division Error;
- OverFlow Error;
- Floating Point Error.

All of the above exceptions fall under the arithmetic base class and are raised for errors in arithmetic operations, as discussed.

Zero Division. When the divisor (second argument of the division) or the denominator is zero, the result raises a zero-division error.

```
try:
    a = 100 / 0
    print (a)
except ZeroDivisionError:
        print ("Zero Division Exception Raised." )
else:
    print ("Success, no error!")
Zero Division Exception Raised.
```

OverFlow Error. This is raised when the result of an arithmetic operation is out of range. OverflowError is raised for integers that are outside a required range.

```
try:
     import math
     print(math.exp(1000))
except OverflowError:
        print ("OverFlow Exception Raised.")
else:
     print ("Success, no error!")
OverFlow Exception Raised.
```

Assertion Error. When an assert statement is failed, an Assertion Error is raised. Let's take an example to understand this. Let's say you have two variables a and b, which you need to compare. To check whether a and b are equal or not, you apply an assert keyword before that, which will raise an Assertion exception when the expression returns false.

```
try:
     a = 100
     b = "DataCamp"
     assert a == b
except AssertionError:
        print ("Assertion Exception Raised.")
else:
     print ("Success, no error!")
Assertion Exception Raised.
```

Attribute Error. When a non-existent attribute is referenced, and when that attribute reference or assignment fails, an attribute error is raised.

In the below example, you can observe that the Attributes class object has no attribute with the name attribute.

```
class Attributes(object):
     a = 2
     print (a)

try:
     object = Attributes()
     print (object.attribute)
except AttributeError:
     print ("Attribute Exception Raised.")
2
Attribute Exception Raised.
```

Import Error. This is raised when you try to import a module that does not exist (is unable to load) in its standard path or even when you make a typo in the module's name.

```
import nibabel
------------------------------------------------------------------
-------------
ModuleNotFoundError                      Traceback (most
recent call last)
<ipython-input-6-9e567e3ae964> in <module>
----> 1 import nibabel

ModuleNotFoundError: No module named 'nibabel'
```

Lookup Error. This acts as a base class for the exceptions that occur when a key or index used on a mapping or sequence of a list/dictionary is invalid or does not exist. The two types of exceptions raised are:

- **Key Error:** If a key you are trying to access is not found in the dictionary, a key error exception is raised.

    ```
    try:
    a = {1:'a', 2:'b', 3:'c'}
    print (a[4])
    except LookupError:
    print ("Key Error Exception Raised.")
    else:
    print ("Success, no error!")
    Key Error Exception Raised.
    ```

- **Index Error:** When you are trying to access an index (sequence) of a list that does not exist in that list or is out of range of that list, an index error is raised.

    ```
    try:
        a = ['a', 'b', 'c']
        print (a[4])
    except LookupError:
        print ("Index Error Exception Raised, list index
    out of range")
    else:
        print ("Success, no error!")
    Index Error Exception Raised, list index out of range
    ```

Name Error: This is raised when a local or global name is not found. In the below example, the ans variable is not defined. Hence, you will get a name error.

```
try:
    print (ans)
except NameError:
    print ("NameError: name 'ans' is not defined")
else:
    print ("Success, no error!")
NameError: name 'ans' is not defined
```

Runtime Error. This acts as a base class for the NotImplemented Error. Abstract methods in user-defined classes should raise this exception when the derived classes override the method.

```
class BaseClass(object):
    """Defines the interface"""
    def __init__(self):
        super(BaseClass, self).__init__()
    def do_something(self):
        """The interface, not implemented"""
        raise NotImplementedError(self.__class__.__name__ +
'.do_something')

class SubClass(BaseClass):
    """Implementes the interface"""
    def do_something(self):
        """really does something"""
        print (self.__class__.__name__ + ' doing something!')

SubClass().do_something()
BaseClass().do_something()
SubClass doing something!
```

Type Error. This exception is raised when two different or unrelated types of operands or objects are combined. In the below example, an integer and a string are added, which results in a type error:

```
try:
    a = 5
    b = "DataCamp"
    c = a + b
except TypeError:
    print ('TypeError Exception Raised')
else:
    print ('Success, no error!')
TypeError Exception Raised
```

Value Error. This is raised when the built-in operation or function receives an argument that has a correct type but invalid value. In the below example, the built-in operation float receives an argument, which is a sequence of characters (value), which is invalid for a type float.

```
try:
    print (float('DataCamp'))
except ValueError:
    print ('ValueError: could not convert string to float:
\'DataCamp\'')
else:
    print ('Success, no error!')
ValueError: could not convert string to float: 'DataCamp'
```

Python Custom Exceptions. Python has many built-in exceptions that you can use in your program. Still, sometimes, you may need to create custom exceptions with custom messages to serve your purpose. You can achieve this by creating a new class, which will be derived from the pre-defined Exception class in Python.

```
class UnAcceptedValueError(Exception):
    def __init__(self, data):
        self.data = data
    def __str__(self):
        return repr(self.data)

Total_Marks = int(input("Enter Total Marks Scored: "))
try:
    Num_of_Sections = int(input("Enter Num of Sections: "))
    if(Num_of_Sections < 1):
        raise UnAcceptedValueError("Number of Sections can't
be less than 1")
except UnAcceptedValueError as e:
    print ("Received error:", e.data)
Enter Total Marks Scored: 10
Enter Num of Sections: 0
Received error: Number of Sections can't be less than 1
```

In the above example, you can see that if you enter anything less than 1, a custom exception will be raised and handled. Many standard modules define their exceptions to report errors that may occur in functions they define.

Demerits of Python exception handling

Making use of Python exception handling has a side effect, as well. Programs that make use of `try-except` blocks to handle exceptions run slightly slower, and the size of your code increases.

Below is an example where the `timeit` module of Python is used to check the execution time of two different statements. In `stmt1`, `try-except` is used to handle ZeroDivisionError, while in `stmt2`, an `if` statement is used as a normal check condition. Then, you execute these statements 10,000 times with variable `a=0`. The point to note here is that the execution time of both the statements is different. You will find that `stmt1`, which handles the exception, took a slightly longer time than `stmt2`, which checks the value and does nothing if the condition is not met [110].

Hence, you should limit the use of Python exception handling and use it for rare cases only. For example, when you are not sure whether the input will be an integer or a float for arithmetic calculations or are not sure about the existence of a file while trying to open it.

```
import timeit
setup="a=0"
stmt1 = '''\
try:
    b=10/a
except ZeroDivisionError:
    pass'''
stmt2 = '''\
if a!=0:
    b=10/a'''
print("time=",timeit.timeit(stmt1,setup,number=10000))
print("time=",timeit.timeit(stmt2,setup,number=10000))
time= 0.003897680000136461
time= 0.0002797570000439009
```

13.3 Common Gateway Interfaces

The word CGI is the acronyms of "Common Gateway Interface," which is used to define how information is exchanged between the webserver and a custom script. The National Center for Supercomputing Applications (NCSA) officially manages CGI scripts.

The CGI is a standard for external gateway programs to interface with servers, such as HTTP servers.

In simple words, it is the collection of methods used to set up a dynamic interaction between the server and the client application. When a client sends a request to the web server, the CGI programs execute that particular request and send back the result.

The users may submit the information in a web browser by using an HTML <form> or <isindex> element. Servers has a special directory called cgi-bin, where CGI script is generally stored. When a client makes a request, the server adds the additional information to the request.

This additional information can be the hostname of the client, the query string, the requested URL, and so on. The web server executes and sends the output to the web browser (or other client application).

Python provides the CGI module, which helps to debug the script, and also the support for the uploading files through an HTML form.

So here the question arises: what does a Python CGI script output look like? The HTTP's server returns the output as two sections separated by a blank line. The first section grasps the number of headers, notifying the client as to what kind of data follows.

The following example shows the generation of the minimal header section in Python CGI programming.

1. # HTML follows
2. print("Content-Type: text/html")
3. # blank line, end of headers
4. print()

The first statement states that HTML code follows; the blank line indicates the header is ended. Let's look at another example.

1. print("<title> This is a CGI script output</title>")
2. print("<h1>This is my first CGI script</h1>")
3. print("Hello, JavaTpoint!")

Web browsing

Before understanding CGI concepts, we need to know the internal process of a webpage or URL when we click on the given link:

- The client (web browser) communicates with the HTTP server and asks for the URL, i.e. the filename.
- If the web browser finds that requested file, then it sends it to the client (web browser), otherwise it sends an error message to the client as an error file.
- The web browser displays either the received file or an error message.

However, we can set an HTTP server so that whenever the user requests a particular dictionary, then it is sent to the client; otherwise, it is executed as a program and whatever the result is it is sent back to the client to be displayed. This process is called the CGI; the programs are called CGI scripts. We can write CGI programs as Python Script, PERL, Script, Shell Script, C or C++ and so on [111].

Configure an Apache webserver for CGI

Python provides the CGI module, which consists of many useful built-in functions. We can use these by importing the CGI module:

- import cgi

Now, we can write further script:

- import cgi
- cgitb.enable()

The above script will cause an exception handler to show a detailed report in the web browser of the errors that have occurred. We can also save the report by using the following script:

- import cgitb
- cgitb.enable(display=0, logdir="/path/to/logdir")

The above feature of the CGI module is helpful during script development. These reports help us to debug the script effectively. When we get the expected output, we can remove them.

Previously, we have discussed users saving information using a form. So how can we get that information? Python provides the FieldStorage class. We can apply the encoding keyword parameter to the document if the form contains the non-ASCII character. We will find the content <META> tag in the <HEAD> section in our HTML document.

The FieldStorage class reads the form's information from the standard input or the environment.

A FieldStorage instance is the same as the Python dictionary. We can use len() and all the dictionary functions in the FieldStorage instance. This looks over the fields with empty string values. We can also consider the empty values using the optional keyword parameter keep_blank_values by setting True.

Example:

```
form = cgi.FieldStorage()
if ("name" not in form or "addr" not in form):
    print("<H1>Error</H1>")
    print("Please enter the information in the name and
address fields.")
    return
print("<p>name:", form["name"].value)
print("<p>addr:", form["addr"].value)
#Next lines of code will execute here...
```

In the above example, we have used the form ["name"], here the name is key. This is used to extract the value which is entered by the user.

We can use the getvalue() method to fetch the string value directly. This function also takes an optional second argument as a default. If the key is not present, it returns the default value.

If the submitted form's data have more than one field with the same name, we should use the form.getlist() function. This returns the list of strings. Looking at the following code, we can add any number of username fields, separated by commas:

```
value1 = form.getlist("username")
usernames1 = ",".join(value)
```

If the field is an uploaded file, then it can be accessed by the value attribute or the getvalue() method and then read that uploaded file in bytes. Let's look at the following code where the user uploads the file:

```
file_item = form["userfile"]
if (fileitem.file):
    # It represent the uploaded file; counting lines
    count_line = 0
    while(True):
        line = fileitem.file.readline()
        if not line: break
        count_line = count_line + 1
```

Sometimes an error can interrupt the program while reading the content of the uploaded file (when the user clicks on the cancel button or back button). The FieldStorage class provides the done attribute which is set to the value -1.

If we submit the form in the "old" format, the item will be an instance of the class MiniFieldStorage. In this class, the list, file and filename attributes are always None.

Generally, the form is submitted via POST and contains a query string with both FieldStorage and MiniStorage items.

The FieldStorage instance uses the many built-in methods to manipulate the user's data. Below are a few of these methods.

First CGI program

We have created a new folder called example in xampp's htdocs folder. Then, we write a Python script, which includes the HTML tags. Let's look at the following directory structure and the demo.py file:

demo.py

1. print ("Content-Type: text/html\r\n\r\n")
2. # then come the rest hyper-text documents
3. print ("<html>")
4. print ("<head>")
5. print ("<title>My First CGI-Program </title>")
6. print ("<head>")
7. print ("<body>")
8. print (" <h1>This is my CGI script </h1> ")
9. print ("</body>")
10. print ("</html>")

Its directory structure is as follows.

Type the localhost/example/demo.py into the web browser. It will display the following output.

Structure of a Python CGI program

Let's look at the following structure of a program:

- The CGI script must contain two sections, separated by a blank line.
- The header must be in the first section, and the second section will contain the kind of data that will be used during the execution of the script.

When scripting a CGI program in Python, take note of the following commonly used syntaxes.

HTML header

In the above program, the line Content-type:text/html\r\n\r\n is a portion of the HTTP, which we will use in our CGI programs.

```
HTTP Field Name: Field Content
```

For example:

```
Content-type: text/html\r\n\r\n
```

TABLE 13.1

File Syntax and Description

Sr.	Header	Description
1.	Content-type	This is a Multipurpose Internet Mail Extension (MIME) string that is used to define the format of the file being returned.
2.	Expires: Date	This displays the valid date.
3.	Location: URL	This is the URL that is returned by the server.
4.	Last-modified: Date	This displays the date of the last modification of the resource.
5.	Content-length: N	This information is used to report the estimated download time for a file.
6.	Set-Cookies: String	This is used to set the cookies by using `string`.

CGI environment variables

We should remember the following CGI environment variable along with the HTML syntax. Let's look at the commonly used CGI environment variables [112].

- **CONTENT_TYPE** This describes the data and type of content.
- **CONTENT_LENGTH** This defines the length of a query or information.
- **HTTP_COOKIE** This is used to return the cookie, which is set by the user in the current scenario.
- **HTTP_USER_AGENT** This variable is used to display the type of browser that the user is currently using.
- **REMOTE_HOST** This is used to describe the path of the CGI scripts.
- **PATH_INFO** This variable is used to define the path of the CGI script.
- **REMOTE_ADDR** We can define the IP address of the visitor by using it.
- **REQUEST_METHOD** This is used to make a request either via POST or GET.

Functions of Python CGI programming

The CGI module provides the many functions that work with it. We now define a few of the important functions:

- **parse(fp = None, environ = os.environ, keep_blanks_values = False, strict_parsing = False)** This is used to parse a query in the environment. We can also parse it using a file, the default for which is **sys.stdin**.

- **parse_qs(qs, keep_blank_values = False, strict _parsing = False)**
  While this is DE integrated, Python uses it for urllib.parse.parse_qs() instead.
- **parse_qsl(qs, keep_blank_value = False, strict_parsing = False)**
  This is also DE integrated, and maintains backward-compatibility.
- **parse_multipart(fb, pdict)** This is used to parse input of type multi-part/form-data for file uploads. The first argument is the **input file**, and the second argument is a dictionary holding the other parameters in the content-type header.
- **parse_header(string)** This is used to parse the header. It permits the MIME header into the main value and a dictionary of parameters.
- **test()** This is used to test a CGI script, which we can use in our program. It will generally write minimal HTTP headers.
- **print_form(form)** This formats a form in HTML.
- **print_directory()** This formats the current directory in HTML.
- **escape(s, quote = False)** The **escape()** function is used to convert characters '<', '>' and '&' in strings into an HTML safe sequence.

Debugging CGI scripts

First, we need to check the trivial installation error. Most of the time, the error occurs during the installation of the CGI script. At the start, we must follow the installation instructions and try installing a copy of the module file `cgi.py` as a CGI script.

Next, we can use the `test()` function from the script. Type the following code with a single statement:

- cgi.test()

Advantages of CGI programming

There are various advantages to using CGI programming:

- They are language independent. We can use them with any programming language.
- CGI programs can work on almost any web server and are portable.
- CGI programs can perform both simple and complex tasks, which means they are fairly scalable.
- CGIs can increase dynamic communication in web applications.
- CGIs can also be profitable, if we use them in development, as they reduce development costs and maintenance costs.
- CGIs takes less time to process requests.

Disadvantages of CGI

The disadvantages of CGI are:

- CGI programs are too complex and hard to debug.
- When we initiate the program, the interpreter has to evaluate a CGI script in each initiation. The result is that this creates a lot of traffic because there are many requests from the side of the client-server.
- CGI programs are quite vulnerable, as most of them are free and easily available without server security.
- CGI uses a lot of processing time.
- During page load, the data aren't stored in cache memory.
- There are huge extensive codebases, most of it in Perl.

Common problems and solutions

Problems can occur during implementation of the CGI script on the server. Here are a few common problems and their solutions:

- First of all, check the installation instructions. Most of the problems occur during the installation. Follow the installation guide properly.
- Check the HTTP server's log file. The `tail -f logfile` in a separated window may be valuable.
- In the CGI, it is possible to display the progress report on the client's screen of running requests. Most HTTP servers save the output from the CGI script until the script is finished.
- Before executing the file, check the syntax error in your script, following `python script.py`.
- If the script does not have any syntax errors then import the library, such as `import cgitb; cgitb.enable()`, to the top of the script.
- The absolute path must be included when importing the external program. The path is usually not set to a very useful value in a CGI script.
- When reading or writing external files, make sure that they can be read or written by the user under which your CGI script will be running. This is an authorized user ID where the script file that the web server is running or some specified user ID for a web server.
- It should be remembered that the CGI script must not set in `setuid`. It won't work on most systems, and is also a security liability.

13.4 Database Access in Python

The Python database interface specification is Python DB-API. This specification is adopted by other Python database interfaces [114].

For your case, you should pick the correct site. A broad variety of servers such as Python Database API are provided by:

- GadFly database servers;
- mSQL database servers;
- MySQL database servers;
- PostgreSQL database servers;
- Microsoft SQL Server 2000 database servers;
- Informix database servers;
- Interbase database servers;
- Oracle database servers;
- Sybase database servers.

Python applications available are Python Database Applications and Application Programmer Interfaces (APIs). For each database you need to use, you may download a separate Mysql API plugin. You have to install both Oracle and MySQL database packages, for example if you need to use an Oracle database and a MySQL database.

The DB API offers where possible a basic standard for using Python constructs and syntax for dealing with databases:

- API module import;
- Acquisition of a database link;
- SQL statements and saved proceedings are released;
- Link closing.

Both principles will be learned using MySQL, so let's discuss the MySQLdb framework.

13.4.1 What is MySQLdb?

MySQLdb is a Python database server communication interface. It is developed over the MySQL C API, which contains the Python Database API v2.0.

How do I install MySQLdb?

You check that you have installed MySQLdb on your device before continuing. You only key in and run the following Python script

Connecting MySQL with Python

To create a connection between the MySQL database and Python, the connect() method of mysql.connector module is used. We pass the database details like HostName, username, and the password in the method call, and then the method returns the connection object.

The following steps are required to connect SQL with Python:

Step 1: Download and Install the free MySQL database from here.

Step 2: After installing the MySQL database, open your Command prompt.

Step 3: Navigate your Command prompt to the location of PIP. Click here to see, How to install PIP?

Step 4: Now run the commands given below to download and install "MySQL Connector". Here, mysql.connector statement will help you to communicate with the MySQL database.

Download and install "MySQL Connector"
pip install mysql-connector-python

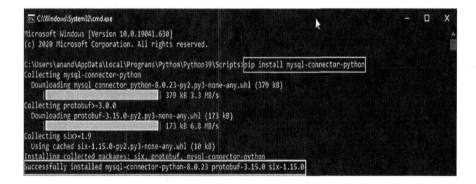

Step 5: Test MySQL Connector

To check if the installation was successful, or if you already installed "MySQL Connector", go to your IDE and run the given below code :

```
import mysql.connector
```

```
Python 3.9.0 Shell
File  Edit  Shell  Debug  Options  Window  Help
Python 3.9.0 (tags/v3.9.0:9cf6752, Oct  5 2020, 15:34:40) [MSC v
27 64 bit (AMD64)] on win32
Type "help", "copyright", "credits" or "license()" for more info
tion.
>>> import mysql.connector
>>>
```

13.4.2 Database Connection

Step 6: Create Connection
Now to connect SQL with Python, run the code given below in your IDE.

```
# Importing module
import mysql.connector

# Creating connection object
mydb = mysql.connector.connect(
        host = "localhost",
        user = "yourusername",
        password = "your_password"
)

# Printing the connection object
    print(mydb)
```

```
python practice1
C:\Users\Anaconda3\python.exe "D:/PYTHON/python practice1.py"
<mysql.connector.connection_cext.CMySQLConnection object at 0x0000013C815F14C8>
Process finished with exit code 0
```

During the execution of the script, my Linux machine shows the following output:

Database version: 5.0.45

In relation to the data, the frame is created and a connection object will be returned and stored for use in db; otherwise, db will be set to zero. Next to the database object, a cursor object will be created and SQL queries will, in turn, be executed. Finally, this assures the connection to the network and the availability of services when it comes online [113].

Creating a Database Table

To create a database, we will use CREATE DATABASE database_name statement and we will execute this statement by creating an instance of the 'cursor' class.

```python
import mysql.connector

mydb = mysql.connector.connect(
        host = "localhost",
        user = "yourusername",
        password = "your_password"
)

# Creating an instance of 'cursor' class
# which is used to execute the 'SQL'
# statements in 'Python'
cursor = mydb.cursor()

# Creating a database with a name
#  execute() method
# is used to compile a SQL statement
# below statement is used to create
# the database
cursor.execute("CREATE DATABASE")
```

The INSERT operation

Records should be produced in a database table. Example: Run the SQL INSERT statement to construct a record in the EMPLOYEE table:

```
import MySQLdb

# Open database connection
db = MySQLdb.connect("localhost", "testuser", "test123",
"TESTDB")

# prepare a cursor object using cursor() method
cursor = db.cursor()

# Prepare SQL query to INSERT a record into the database.
sql = """INSERT INTO EXPLOYEE(FIRST_NAME, LAST_NAME,
      AGE, SEX, INCOME) VALUES ('Mac', 'Mohan', 20, 'M',
      2000)"""
try:
  #Execute the SQL command
  cursor.execute(sql)
  #Commit your changes in the database
  db.commit()
except:
  #Rollback in case there is any error
  db.rollback()

#disconnect from server
db.close()
```

The example above can be written to dynamically generate SQL queries:

```
import MySQLdb
# Open database connection
db = MySQLdb.connect("localhost", "testuser", "test123",
"TESTDB")

# prepare a cursor object using cursor() method
cursor = db.cursor()

# Prepare SQL query to INSERT a record into the database.
sql = "INSERT INTO EXPLOYEE(FIRST_NAME, \LAST_NAME, AGE,
      SEX, INCOME)\ VALUES ('%s', '%s', '%d', '%c', '%d')" %
      \('Mac', 'Mohan', 20, 'M', 2000)
try:
  #Execute the SQL command
  cursor.execute(sql)
```

```
    #Commit your changes in the database
    db.commit()
except:
    #Rollback in case there is any error
    db.rollback()

#disconnect from server
db.close()"
```

Another example of an execution type is the code section where parameters can be transferred directly:

```
.......................................
user_id = "test123"
password = "password"

con.execute('insert into Login values("%s", "%s")'%\
    (user_id, password))
.......................................
```

13.5 The Read Operation

Some database processes involve gathering valuable database information. After you have developed a database link, you are prepared to access the database. Use either fetchone () for fetch calling a single record or fetch-all () for fetching from a table of database multiple records [115]:

- **fetchone()** This retrieves a question response collection from the next page. A collection of results is an object returned to query a table using a cursor variable.
- **fetchall()** The result collection lists all access. If any lines of the result set were already discarded, the remainder of the lines of the result set was retried.
- **rowcount** This is a read-only function which returns an execute() method to the number of rows affected.

For example, the following protocol checks all EMPLOYEE table documents with pay above 1000:

```
import MySQLdb

# Open database connection
db = MySQLdb.connect ("localhost", "testuser", "test123",
"TESTDB")

# prepare a cursor object using cursor () method
cursor = db.cursor()

sql = "SELECT * FROM EMPLOYEE\
      WHERE INCOME > '%d'" % (1000)
try:
   # Execute the SQL command
   cursor.execute (sql)
   # Fetch all the rows in a list of lists.
   results =  cursor.fetchall()
   for row in results:
       fname = row[0]
       lname = row[1]
       age = row[2]
       sex = row[3]
       # Now print fetched result
       print "fname=%s, lname=%s, age=%d, sex=%s, income=%d" %\
             (fname, lname, age, sex, income)
except:
   print "Error: unable to fetch data"

# disconnect form server
db.close() "
```

This will produce the following result:

Update operation

The Check Any storage operation ensures that one or more documents currently in storage can be updated. The process below changes all SEX records to "M." Here, the AGE of all men is raised by one year.

Example:

```python
import MySQLdb

# Open database connection
db = MySQLdb.connect ("localhost", "testuser", "test123",
"TESTDB")

# prepare a cursor object using cursor () method
cursor = db.cursor ()

# Prepare SQL query to UPDATE required records
sql = "UPDATE EMPLOYEE SET AGE = AGE + 1
                              WHERE SEX = '%c'" % ('M')
try:
    # Execute the SQL command
    cursor.execute (sql)
    # Commit your changes in the database
    db.commit ()
except:
    # Rollback in case there is any error
    db.rollback ()

# disconnect form server
db.close () "
```

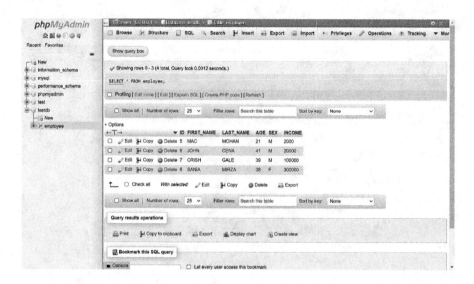

DELETE operation

If you want to remove records from your archive, a remove operation is needed. Action is then taken to erase all EMPLOYEE documents in which AGE is higher than 20.

Example:

```
import MySQLdb
# Open database connection
db = MySQLdb.connect ("localhost", "testuser", "test123",
"TESTDB")

# prepare a cursor object using cursor () method
cursor = db.cursor()

# Prepare SQL query to DELETE required records
sql = "DELETE FROM EMPLOYEE WHERE AGE > '%d'" % (20)
try:
    # Execute the SQL command
    cursor.execute (sql)
    # Commit your changes in the database
    db.commit()
except:
    # Rollback in case there is any error
    db.rollback()

# disconnect form server
db.close() "
```

Performing transactions

Transactions are a method for maintaining accuracy in results. The following four principles relate to transactions:

- **Atomicity transactions**: If a transaction is ended or nothing is done.
- **Consistency transactions**: In a consistent state, a process must start and exit the device.
- **Isolation transactions**: Except for the current transaction, intermediate contract outcomes are not available.
- **Durability transactions**: The consequences are permanent until an operation is committed, even though a system malfunction is involved.

The Python DB API 2.0 provides two methods to either *commit* or *rollback* a transaction. You already know how to implement transactions. Here again is a similar example:

```
# Prepare SQL query to DELETE required records
sql = "DELETE FROM EMPLOYEE WHERE AGE > '%d'" % (20)
try:
  # Execute the SQL command
  cursor.execute (sql)
  # Commit your changes in the database
  db.commit()
except:
  # Rollback in case there is any error
  db.rollback()
```

COMMIT operation

Commit is a phase, which provides the database with a green light to finalize changes, and after that step, no shift can be changed. This is a clear example of calling commits.db.

```
commit().
```

ROLLBACK operation

Use the `rollback()` mechanism if you are not satisfied with one or more improvements and want to undo them entirely. A basic explanation for calling the `rollback ()` method is:

```
db.rollback()
Disconnecting Database
To disconnect Database connection, use close() method.
db.close()
```

When the user closes the connection to a database using the `close()` process, any remaining transactions are re-rolled by the server. However, it is easier to call commit or rollback directly rather than relying on the lower DB implementation information.

Steps to connect Python to MS Access using pyodbc

Step 1: Install the pyodbc package

To start, install the pyodbc package that will be used to connect Python to Access. You may use PIP to install the pyodbc package:

```
pip install pyodbc
```

Tip: Before you connect Python to Access, you may want to check that your Python bit version matches with your MS Access bit version (e.g. use Python 64 bit with MS Access 64 bit).

Step 2: Create the database and table in Access

Next, let's create:

- An Access database called: **test_database**
- A table called: **products**

- The products table would contain the following columns and data:

product_id	product_name	price
1	Computer	800
2	Printer	150
3	Desk	400
4	Chair	120
5	Tablet	300

Step 3: Connect Python to Access

To connect Python to Access:

- Add the path where you stored the Access file (after the syntax **DBQ=**). Don't forget to add the MS Access file extension at the end of the path ('accdb').
- Add the table name within the **select** statement.

```
import pyodbc

conn = pyodbc.connect(r'Driver={Microsoft Access Driver
(*.mdb, *.accdb)};DBQ=path where you stored the Access
file\file name.accdb;')
cursor = conn.cursor()
cursor.execute('select * from table_name')

for row in cursor.fetchall():
  print (row)
```

For example, let's suppose that the Access database is stored under the following path:

```
C:\Users\Ron\Desktop\Test\test_database.accdb
```

where **test_database** is the MS Access file name within that path, and **accdb** is the MS Access file extension.

Before you run the code below, you'll need to adjust the path to reflect the location where the Access file is stored on *your* computer (also don't forget to specify the table name within the **select** statement. Here, the table name is *products*):

```
import pyodbc

conn = pyodbc.connect(r'Driver={Microsoft Access Driver
(*.mdb, *.accdb)};DBQ=C:\Users\Ron\Desktop\Test\test_
database.accdb;')
cursor = conn.cursor()
```

```
cursor.execute('select * from products')

for row in cursor.fetchall():
    print (row)
```

Step 4: Run the code in Python

Run the code in Python, and you'll get the same records as stored in the Access table:

```
(1, 'Computer', 800)
(2, 'Printer', 150)
(3, 'Desk', 400)
(4, 'Chair', 120)
(5, 'Tablet', 300)
```

Handling errors

Many sources of errors will remain. Examples include an executed SQL command syntax error, a connectivity malfunction or a call to the fetch protocol to handle a cancelled or completed sentence. In database Module the MySQL API describes a list of errors. These cases are specified in Table 13.2.

TABLE 13.2

Error Handling

Sr.No.	Exception and Description
1	**Warning** Used by fatal problems. Does StandardError have to be subset.
2	**Error** Error Basic Class. The StandardError Subclass Must.
3	**InterfaceError** It is not the database itself used for mistakes in the database module. It is essential to subclass error.
4	**DatabaseError** Used in the database for bugs. Must Error subclass.
5	**DataError** DatabaseError subclass which refers to data errors.
6	**OperationalError** DatabaseError subclass which refers to errors like losing a database link. In general, these errors are beyond Python's command.
7	**IntegrityError** Database Error Subclass for conditions damaging relationship integrity, such as restrictions on individuality or foreign keys.

(Continued)

TABLE 13.2 (CONTINUED)

Sr.No.	Exception and Description
8	**InternalError** The DatabaseError subclass corresponds to internal errors of the database component, such as the non active cursor.
9	**ProgrammingError** DatabaseError Subclass refers to errors like a wrong table name.
10	**NotSupportedError** The DatabaseError subclass is intended to attempt to call unsupported functions.

13.6 Python Multithreaded Programming

Many threads operating simultaneously are equivalent to multiple programs, but with the following advantages:

- Several process threads occupy the same disk space with the original post and thus more effectively bind or communicate with each other than when different processes are involved.
- Threads are often considered lightweight and use lower power overheads in operations.

A thread has a start, an execution and an end. It has a command indicator to show where it is running, within its context:

- Preemption (interruption) is possible.
- It can be temporarily halted during other lines, also known as dreaming, and is called yielding.

13.6.1 Starting a New Thread

You need to call the `thread` function in the class module to construct another class:

```
thread.start_new_thread ( function, args[, kwargs] )
```

This call method helps you to build new threads easily and efficiently in Linux and Windows.

The process call comes back right away, and the child thread begins and calls with the passed `args` array. The loop ends when the method returns. `Args` is here a multitude of arguments; use an empty tuple without arguments to name the method. `kwargs` is a keyword statement conditional dictionary [116].

Example:

```
import thread
import time

# Define a function for the thread
def print_time (threadName, delay):
    count = 0
    while count < 5:
        time.sleep (delay)
        count += 1
        print ("%s: %s" % (threadName, time.ctime(time.time())))

#Create two threads as follows
try:
    thread.start_new_thread(print_time, ("Thread-1", 2, ))
    thread.start_new_thread(print_time, ("Thread-1", 4, ))
except:
    print "Error: unable to start thread"

while 1:
    pass
```

When the above code is executed, it produces the following result:

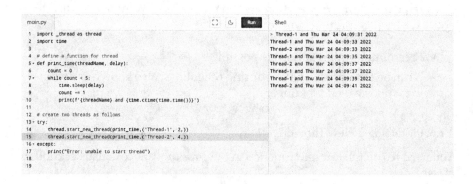

This is very good for low-level threading, but relative to the current thread module, the thread module has very low power.

13.6.2 The Threading Module

The newest connector framework in Python 2.4 delivers much better, high-level thread support than the previous portion of the connector framework.

The threading module displays all threading module methods and includes some supplementary methods:

- **threading.activeCount()** Returns the amount of active threads.
- **threading.currentThread()** The number of things returned in the caller thread control.
- **threading.enumerate()** Returns a list of all thread objects currently in use.
- A thread class also includes the threading module, which implements the thread. Besides the procedures as a method of the thread.
- **run()Thread Method** The function run() is the thread input stage.
- **start() Thread Method** The start() process begins with an execution method call for a thread.
- **join([time]) Thread Method** The join() waits for threads to terminate.
- **isAlive() Thread Method** The function isAlive() tests whether a thread continues to run.
- **getName()Thread Method** The function getName() returns a thread's name.
- **setName() Thread Method** The function setName() sets the thread's name.

13.6.3 Thread Module

You must do the following to add a current module thread.

- Create a new subclass of thread hierarchy.
- Bypass the init process (self, [args]) to add extra arguments.
- Then override the run (self, [args]) function under which the loop is introduced at the top.

If the new thread subclass has been developed, you may build an example and launch the new thread with the invoking begin() process.

Example:

```
import threading
import time

exitFlag = 0

class myThread(threading.Thread):
    def__init__(self, threadID, name, counter):
        threading.Thread.__init__(self)
        self.threadID = threadID
        self.name = name
        self.counter = counter
```

```
    def run(self):
        print "Starting" + self.name
        print_time(self.name, 5, self.counter)
        print "Exiting" + self.name

def print_time(threadName, counter, delay):
    while counter:
        if exitFlag:
            threadName.exit()
        time.sleep(delay)
        print "%s: %s" %(threadName, time.ctime(time.time()))
        counter -= 1

# Create new threads
thread1 = myThread(1, "Thread-1", 1)
thread2 = myThread(2, "Thread-2", 2)

# Start new Threads
thread1.start()
thread2.start()

print "Exiting Main Thread"
```

When the above code is executed, it produces the following result:

```
Starting <bound method Thread.start of <myThread(Thread-1,
Started 139990751456032)>>
Starting <bound method Thread.start of <myThread(Thread-2,
Started 139990750133024)>>
Exiting Main Thread
Thread-1 and Thu Mar 24 04: 27: 40 2022
Thread-2 and Thu Mar 24 04: 27: 41 2022
Thread-1 and Thu Mar 24 04: 27: 41 2022
Thread-1 and Thu Mar 24 04: 27: 42 2022
Thread-2 and Thu Mar 24 04: 27: 43 2022
Thread-1 and Thu Mar 24 04: 27: 43 2022
Thread-1 and Thu Mar 24 04: 27: 44 2022
Starting <bound method Thread.start of <myThread(Thread-1,
Started 139990751456032)>>
Thread-2 and Thu Mar 24 04: 27: 45 2022
Thread-2 and Thu Mar 24 04: 27: 47 2022
Thread-2 and Thu Mar 24 04: 27: 49 2022
Starting <bound method Thread.start of <myThread(Thread-2,
Started 139990750133024)>>

**Process exited - Return Code: 0**
Press Enter to exit terminal
```

13.6.4 Priority Multithreaded Queue

You may build a new queue object with a certain number of things in the queue module. The queue is managed using the following methods:

- **get() priority queue** The get() removes and returns an item from the queue.
- **put() priority queue** Adds in queue only.
- **qsize() priority queue** The quantity of products in the queue will be returned.
- **empty() priority queue** Returns empty() True if the queue is blank; False if not.
- **full() priority queue** Returns complete() True if the queue is complete; False otherwise.

13.7 Networking in Python

Two tiers of network connectivity control are supported by Python. At a low level, you can access the underlying operating system's simple socket support which helps you to introduce connecting protocol clients and servers. Python also has libraries to offer higher levels of connectivity for programs such as FTP and HTTP to different network protocols. This section provides you with an interpretation of Network-Socket Programming's popular definition.

13.7.1 What Are Sockets?

Sockets are the endpoints of a two-way system of contact. In a method, sockets can interact on the same computer between processes and various continents. A variety of channel types can be installed with sockets: Unix domain sockets, TCP, UDP, and so on. In addition to the classrooms accessible in the socket library, there is a standardized interface for managing other transportations. Sockets have a language of their own.

13.7.2 The Socket Module

In the socket module with the standard syntax you can use the `socket.socket()` function to create a socket:

```
s = socket.socket (socket_family, socket_type, protocol=0)
```

TABLE 13.3

Sockets

Sr. No.	Terms and Meaning
1	**Domain** As transport mechanism, the protocol family is used. The values of AF INET, PF INET, PF UNIX, PF X25 and so forth are constants
2	**type** The method of contact between the two endpoints, usually SOCK STREAM and SOCK DGRAM for connection-oriented protocols.
3	**protocol** This can usually be used to define a protocol variant within a domain and form
4	**hostname** Network Interface Identifier • A string, that can be a hostname, a dotted Quad address, or a colon (and possibly dot) IPV6 address. • A "<broadcast>" string that indicated an INADDR BROADCAST address. • A null string that specifies the INADDR ANY value, or • A binary address in the byte order of a host.
5	**port** Each server listening on one or more ports for clients calling. A port can be a Fixnum port number, a port number series, or a service name.

This is the function definition:

- **socket_family** As mentioned earlier, this is either AF UNIX or AF INET.
- **socket_type** This is SOCK STREAM or SOCK DGRAM.
- **protocol** Typically, this is 0 by nature.

Until you have an object socket, you can use the functions required to create your client or server applications.

TABLE 13.4

Server Socket Methods

Sr.No.	Methods and Meaning
1	**s.bind()** The hostname, port number pair, is linked to the socket using this form.
2	**s.listen()** This method creates a Transmission Control Protocol (TCP) listener and starts it.
3	**s.accept()** This recognizes the TCP client connection actively, hoping for a connection (blocking).

TABLE 13.5

Client Socket Methods

Sr.No.	Methods and Meaning
1	**s.connect()** This way, the TCP server link is actively initiated.

TABLE 13.6

Socket Methods

Sr.No.	Methods and Meaning
1	**s.recv()** This method gets a response from the TCP.
2	**s.send()** This way the TCP message is sent.
3	**s.recvfrom()** The message for this process is a User Datagram Protocol (UDP).
4	**s.sendto()** UDP message is sent via this process.
5	**s.close()** The socket is closed.
6	**socket.gethostname()** The hostname returns.

A normal server

In order to create a socket object for writing an internet server, we use the socket functions in the socket module. A socket object can be used for other operations to set up a socket server. Now, to set the port you want to use for your link, call bind(hostname, port). Call the approval process of the returned object first. This approach waits until a customer connects to the given port and returns a connected entity connection to the given unit.

```
import socket                        # Import socket module

s = socket.socket()                  # Create a socket object
hots = socket.gethostname()          # Get local machine name
port = 12345                         # Reserve a port for your service.
s.bind((host, port))                 # Bind to the port

s.listen(5)                          # Now wait for client connection.
while True:
    c, addr = s.accept()             # Establish connection with client.
    print 'Got conection from', addr
    c.sent('Thank you for connection')
    c.close()                        # Close the connection"
```

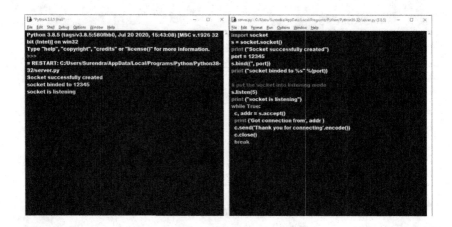

A simple client

Let us write a very simple client program to connect to a specified 12345 Port and server. That's easy to do using Python's module feature to build a socket request.

A TCP connection to the network hostname is opened by `socket.connect(hostname, address)`. When a socket is open, you can read it like any other IO object. When done, remember that you can just close a tab. The following code is a very easy way that connects to a specified host and port, reads and then leaves all the socket data:

```python
import socket               # Import socket module

s = socket.socket()         # Create a socket object
host = socket.gethostname() # Get local machine name
port = 12345                # Reserve a port for your service.

s.connect((host, port))
print s.recv(1024)
s.close()                   # Close the socket when done
```

Now go back to this server.py and then run over client.py to see the answer.

```
# Following would start a server in the background.
$ python server.py &
# Once the server is started run client as follows:
$ python client.py
```

This would produce the following result:

```
got connection from ('127.0.0.1')
thank you for connecting" Python Internet modules
```

TABLE 13.7

Some Important Modules in Python Internet Programming

Protocol	Common function	Port No	Python module
HTTP	Web page	80	httplib, urllib, xmlrpclib
NNTP	Usent news	119	nntplib
FTP	File transfer	20	ftplib, urllib
SMTP	Sending email	25	smtplib
POP3	Fetching email	110	poplib
IMAP4	Fetching email	143	imaplib
Telnet	Comment lines	23	telnetlib
Gopher	Document transfer	70	gopherlib, urllib

13.8 Conclusion

Exception handling helps break the typical control flow of your program by providing a mechanism to decouple Python error handling and makes your code more robust. Python exception handling is one of the prime factors in making your code production-ready and future proof, apart from adding unit testing and object-oriented programming. It is a powerful technique and is a concept of a mere four blocks. try block looks for exceptions thrown by the code, while the except block handles those exceptions (built-in and custom). CGI is a set of standards that defines how information is exchanged between the web server and a custom script. Multiple threads within a process share the same data space with the main thread and can therefore share information or communicate with each other more easily. Python Database

API is the Database interface for standard Python. This standard is adhered to by most Python Database interfaces. There are various Database servers supported by Python Database such as MySQL, GadFly, mSQL, PostgreSQL, Microsoft SQL Server 2000, Informix, Interbase, Oracle and Sybase. To connect with the MySQL database server from Python, we need to import the `mysql.connector` interface.

Case Studies

Python Case Study 1

Backtracking is the approach to a dilemma, based on prior moves. In a labyrinth problem, for example, the strategy relies on all the steps you take. If one such move is false, then the answer is not going to be found. We pick a direction in a labyrinth dilemma and keep going forward. But when we know that the course is false, then we just return and change it. This is simply backtracking.

We take a first step in backtracking to see whether or not this move is right, i.e. whether or not it gives the correct response. But if it doesn't, then we're just returning and changing our first step. This is achieved by recursion in general. So, when creating chances, first we begin with a partial sub-solution (which may lead to or may not lead to a solution) and then investigate whether or not we can continue with this sub-solution. If not, we just return and modify.

The final measures for retrieval are therefore:

- Start a sub-solution;
- Verify whether or not this sub-solution leads to the solution;
- If not, return then change the sub-solution and go on.

N Queens on an N×N Chessboard

Among the most commonly used example is N queens put on an N×N chess game so that no queens will strike down a second queen. The queen can strike in vertical, horizontal or diagonal directions. Similar attempts are also made to solve this dilemma. First, we randomly position the first queen and then put the next queen in a safe spot. We take the following until the number of unranked queens is zero or no safe position is left (a solution is found); see Figure CS.1. If there is no secure place available, then we modify the recently positioned queen's location.

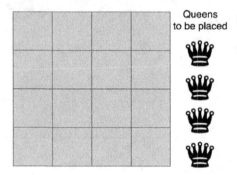

FIGURE CS.1

A chessboard showing where N×N and N queens must be put. So, we're going to proceed with the first queen; see Figure CS.2.

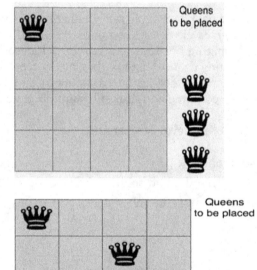

FIGURE CS.2

You will have seen that the last queen can't be positioned in a safe position. We're just going to change the prior queen's position. And that is a reversal. There is still no other position for the third queen so that we can move a step forward and change the second queen's position; see Figure CS.3.

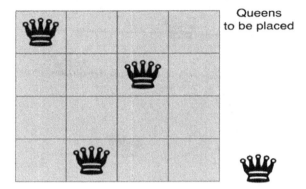

FIGURE CS.2   (CONTINUED)

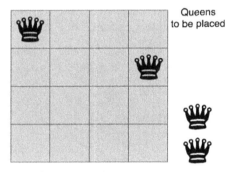

FIGURE CS.3
And now, before we find a way, we can bring the third queen to a secure position again; see Figure CS.4.

FIGURE CS.4
We will start this process and eventually, as seen below, we will find the solution; see Figure CS.5.

FIGURE CS.5

If you now understand backtracking, please let's now use the backtracking form to code the above problem for placing N queens on the NxN chessboard.

Python code

```
#Number of queens
print ("Enter the number of queens"}
N = int (input())

#chessboard
#NxN matrix with all elements 0

board = [[0]*N for _ in range (N)]
def is attack(i, j):
    #checking if there is a queen in row or column
    for k in range(0,N):
        if board[i] [k]==1 or board[k] [j]==1:
            return True
    #checking diagonals
    for k in range(0,N):
        for 1 in range(0,N):
            if (k+1==i+j) or (k-1==i-j):
                if board[k] [1]==1:
                    return True
    return False

def N_queen(n):
    #if n is 0, solution found
    if n==0:
        return True
    for i in range(0,N):
        for j in range(0,N):
            "'checking if we can place a queen here or
            not
```

```
                    queen will not be placed if the place is
                    being attacked
                    or already occupied"'
                    if (not(is_attack(i,j))) and (board[i]
                    [j]!=1}:
                         board[i][j] = 1
                         #recursion
                         #wether we can put the next queen with
                         this arrangment or not
                         if N_queen(n-1)==True:
                              return True
                         board[i][j] = 0

        return False
  N_queen(N)
  for i in board:
    print (i)
```

Explanation of the Code

is_attack(int i,int j) → This is a feature to see if any other queen strikes the cell I j). We only verify if the 'i' or 'j' column has some other queen. We then verify that on the cell diagonal cells I j) there is any queen or not. Any cell (k, l), whether k+l equals i+j or k-l equals I would be diagonal to the cell I j).

N_queen → This is the role under which the monitoring algorithm is actually applied.

if(n==0) → When there is no queen present, that implies that there is a solution for all queens.

if((!is_attack(i,j)) && (board[i][j]!=1)) → We just search whether or not the cell has a queen. Is attack tests whether any other queen and board[i][j] are targeting the cell! = 1 means the cell is empty. If we satisfy these conditions, we may add a queen to the cell – board[i][j]=1.

if(N_queen(n-1)==1) → We're now called the feature again and we're watching the rest of the queens here. If this attribute is not valid (for the rest of the queen), then we change the current movement – board[i][j] = 0, and this time the loop will shift the queen to another position.

Python Case Study 2

Let's mix the minimax and appraisal features we've mastered so far with a good Tic-Tac-Toe AI (Artificial Intelligence) for the perfect game. This AI takes into account all possibilities and makes the right decision.

Finding the Best Move

A new feature named findBestMove() will be added. This method uses mini-max() to compare all possible moves and returns the maximizer's best move to make.

Minimax

Whether the latest change is stronger than the optimal move or not, we use the minimax() function that takes into account all the possibilities the game can take and returns the best value for this move, given that the competitor still plays in optimum terms.

In the minimax() function the code for the maximizer and minimizer is like findBestMove(), the main change being that it would return a number instead of returning a move.

Checking for GameOver State

We use the isMovesLeft() feature to verify that play is over and to make sure no moves are left. It is a simple, easy feature that determines whether or not a move is possible, returning true or false.

Making our AI Smarter

One last step is to make our AI much smarter. Even if the next AI is fine, it will plan moves that lead to a gradual victory or a more rapid defeat. Let's take and describe an example.

Suppose that X will obtain the game from a given board state in two ways.

1. Move **A**: X can win in two moves
2. Move **B**: X can win in four moves

On both moves A and B, our measurement feature returns a value of +10. Even if the A move is easier since a quicker win is expected, often our AI will choose B. We remove the depth value from the evaluated score in order to solve this challenge. This means that if you win, you choose the win that takes fewer moves and you expand the game and take as many moves as possible. The new value would then be calculated by:

- Move **A** will have a value of $+10 - 2 = 8$
- Move **B** will have a value of $+10 - 4 = 6$

Now that A move is higher than move B, our AI can choose move A over move B. The same must be achieved for the minimizing unit. We apply the depth value instead of subtracting it, as the minimizer is always trying to achieve the negative. In or beyond the measurement method, the depth may be subtracted. This works everywhere. Outside of the feature, we may decide to do it as in Figure CS.6.

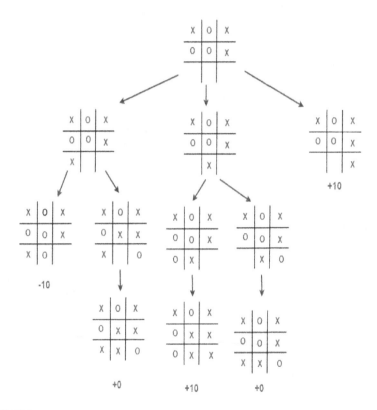

FIGURE CS.6

```
# Tic-Tac-Toe Program using
# random number in Python

# importing all necessary libraries
import numpy as np
import random
from time import sleep

# Creates an empty board
def create board():
```

```
            return(np.array([[0, 0, 0],
                             [0, 0, 0],
                             [0, 0, 0]]))
# Check for empty places on board
def possibilities(board):
        1 = []

        for i in range(len(board)):
                for j in range(len(board)):

                        if board[i][j] == 0:
                                1.append((i, j))
        return (1)

# Select a random place for the player
def random_place(board, player}:
        selection = possibilities(board)
        current_loc = random.choice(selection)
        board[current_loc] = player
        return(board)

# Checks whether the player has three
# of their marks in a horizontal row
def row_win(board, player):
        for x in range(len(board)):
                win = True

                for y in range(len(board)):
                        if board[x, y] != player:
                                win = False
                                continue

                        if win == True:
                        retun(win)
        return(win)

# Checks whether the player has three
# of their marks in a vertical row
def col_win(board, player):
        for x in range(len(board)):
                win = True

                for y in range(len(board)):
                        if board[y][x] != player:
                                win = False continue
                        if win == True:
                                return(win}
        return (win)

# Checks whether the player has three
# of their marks in a diagonal row
def diag_win(board, player):
```

```
                win. = True
                y = 0
                for x in range(len(board)):
                        if board[x, x] != player:
                                win. = False
                if win:
                        return win
                win = True
                if win:
                        for x in range(len(board)):
                                y = len(board) - 1 - x
                                if board[x, y] != player:
                                        win = False
                return win.
# Evaluates whether there is
# a winner or a tie
def evaluate (board.) :
        winner = 0

        for player in [1, 2]:
                if (row_win(board, player) or
                        col_win(board,player) or
                        diag_win(board,player)):

                        winner = player

        if np.all(board != 0) and winner == 0:
                winner = -1
        return winner

# Main function to start the game
def play_game():
        board, winner, counter = create_board(), 0, 1
        print(board)
        sleep(2)

        while winner == 0:
                for player in [1, 2]:
                        board = random_place(board, player)
                        print("Board after" + str(counter)
                        + " move")
                        print(board)
                        sleep (2)
                        counter += 1
                        winner = evaluate(board)
                        if winner != 0:
                                break
        return(winner)

# Driver Code
print("Winner is: " + str(play_game()))
```

Output

```
[[0 0 0]
 [0 0 0]
 [0 0 0]]
Board after 1 move
[[0 0 0]
 [0 0 0]
 [1 0 0]]
Board after 2 move
[[0 0 0]
 [0 2 0]
 [1 0 0]]
Board after 3 move
[[0 1 0]
 [0 2 0]
 [1 0 0]]
Board after 4 move
[[0 1 0]
 [2 2 0]
 [1 0 0]]
Board after 5 move
[[1 1 0]
 [2 2 0]
 [1 0 0]]
Board after 6 move
[[1 1 0]
 [2 2 0]
 [1 2 0]]
Board after 7 move
[[1 1 0]
 [2 2 0]
 [1 2 1]]
Board after 8 move
[[1 1 0]
 [2 2 2]
 [1 2 1]]
Winner is: 2
```

Case Study 3: Eight-puzzle Problem

We are provided with a 3*3 matrix, eight tiles and one empty area (each tile has one to eight numbers). The goal is to arrange the numbers on tiles in the empty space to conform to the final configuration. You float through the open space for four adjacent tiles (left, right, up and down); see Figure CS.7.

Initial configuration   Final configuration

FIGURE CS.7

For example,

1. **Depth First Search (DFS) (Brute-Force)**

   We will scan the state space at depth first (all problem collection, i.e. all statements reachable from the initial state).

   In this strategy, subsequent steps do not get us closer to the target. Regardless of the initial condition, the state space tree quest follows the leftmost route from the center. In this method, a node of response can never be identified; see Figure CS.8

2. **Breadth First Search (BFS) (Brute-Force)**

   We will scan the state space tree for the first time in depth. This is always a target closer to the center. However, whatever the initial condition, the algorithm seeks to shift the same sequence as DFS.

3. **Branch and Bound**

   An "intelligent" rating feature, often an estimated cost feature, will also speed up the discovery of a response node, in order to prevent the quest of substrates that do not have a reaction node. The retrieval technique is similar, but it uses a BFS-like scan.

   In general, there are three types of branch and connected nodes:

   1. A live node is a node formed but not yet generated with children.
   2. An E-node is a node that currently explores its children. That is, an E-node is an increasing node at the moment.
   3. A dead node is a node not to be extended or further discussed.

State Space Tree for 8 Puzzle
Role of cost:
The X node is correlated with a cost in the search tree. The cost function is beneficial for the next E-node. The next E-node is the least costly. The cost function should be specified.

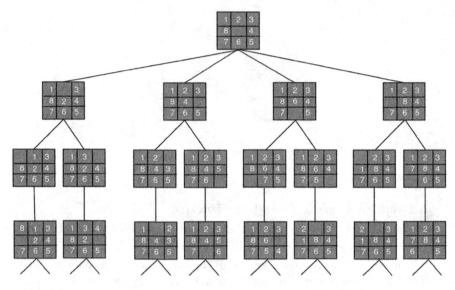

FIGURE CS.8
State space tree for eight puzzle.

```
"C(X) = g(X) + h(X) where
g(X) = cost of reaching the current node
        from the root
h(X) = cost of reaching an answer node from X."
```

Ideal Cost Function for Eight-puzzle Algorithm

We believe that it would cost one unit to shift one tile in either direction. In view of this, for the eight-puzzle algorithm, we define the cost function as follows:

```
"c(x) = f(x) + h(x) where
f(x) is the length of the path from root to x
    (the number of moves so far) and
h(x) is the number of non-blank tiles not in
     their goal position (the number of misplaced tiles).
     There are at least h(x)
     moves to transform state x to a goal state"
```

Program Code

```
"// Program to print path from root node to destination
node
```

```
// for N*N -1 puzzle algorithm using Branch and Bound
// The solution assumes that instance of puzzle is solvable
#include <bits/stdc++.h>
using namespace std;
#define N 3

// state space tree nodes
struct Node
{
    // stores the parent node of the current node
    // helps in tracing path when the answer is found
    Node* parent;

    // stores matrix
    int mat[N][N];

    // stores blank tile coordinates
    int x, y;
    // stores the number of misplaced tiles
    int cost;

    // stores the number of moves so far
    int level;
};
// Function to print N x N matrix
int printMatrix(int mat[N][N])
{
    for (int i = 0; i < N; i++)
    {
        for (int j = 0; j < N; j++)
            printf("%d ", mat[i][j]);
        printf("\n");
    }
}
// Function to allocate a new node
Node* newNode(int mat[N][N], int x, int y, int newX,
        int newY, int level, Node* parent)
{
    Node* node = new Node;

    // set pointer for path to root
    node->parent = parent;

    // copy data from parent node to current node
    memcpy(node->mat, mat, sizeof node->mat);

    // move tile by 1 position
    swap(node->mat[x][y], node->mat[newX][newY]);

    // set number of misplaced tiles
    node->cost = INT_MAX;
```

```
    // set number of moves so far
    node->level = level;

    // update new blank tile cordinates
    node->x = newX;
    node->y = newY;
    return node;
}
// bottom, left, top, right
int row[] = { 1, 0, -1, 0 };
int col[] = { 0, -1, 0, 1 };

// Function to calculate the number of misplaced tiles
// ie. number of non-blank tiles not in their goal position
int calculateCost(int initial[N][N], int final[N][N])
{
    int count = 0;
    for (int i = 0; i < N; i++)
    for (int j = 0; j < N; j++)
        if (initial[i][j] && initial[i][j] != final[i][j])
        count++;
    return count;
}
// Function to check if (x, y) is a valid matrix cordinate
int isSafe(int x, int y)
{
    return (x >= 0 && x < N && y >= 0 && y < N);
}
// print path from root node to destination node
void printPath(Node* root)
{
    if (root == NULL)
        return;
    printPath(root->parent);
    printMatrix(root->mat);

    printf("\n");
}
// Comparison object to be used to order the heap
struct comp
{
    bool operator()(const Node* lhs, const Node* rhs) const
    {
    return (lhs->cost + lhs->level) > (rhs->cost +
    rhs->level);
    }
};
```

```
// Function to solve N*N - 1 puzzle algorithm using
// Branch and Bound. x and y are blank tile coordinates
// in initial state
void solve(int initial[N][N], int x, int y,
    int final[N][N])
{
    // Create a priority queue to store live nodes of
    // search tree;
    priority_queue<Node*, std::vector<Node*>, comp> pq;

    // create a root node and calculate its cost
    Node* root = newNode(initial, x, y, x, y, 0, NULL);
    root->cost = calculateCost(initial, final);

    // Add root to list of live nodes;
    pq.push(root);

    // Finds a live node with least cost,
    // add its childrens to list of live nodes and
    // finally deletes it from the list.
    while (!pq.empty())
    {
        // Find a live node with least estimated cost
        Node* min = pq.top();

        // The found node is deleted from the list of
        // live nodes
        pq.pop();

        // if min is an answer node
        if (min->cost == 0)
        {
            // print the path from root to destination;
            printPath(min);
            return;
        }
        // do for each child of min
        // max 4 children for a node
        for (int i = 0; i < 4; i++)
        {
            if (isSafe(min->x + row[i], min->y + col[i]))
            {
                // create a child node and calculate
                // its cost
                Node* child = newNode(min->mat, min->x,
                    min->y, min->x + row[i],
                    min->y + col[i],
                    min->level + 1, min);
                child->cost = calculateCost(child->mat,
                final);
```

```
                    // Add child to list of live nodes
                    pq.push(child);
                }
            }
        }
}
// Driver code
int main()
{
    // Initial configuration
    // Value 0 is used for empty space
    int initial[N][N] =
    {
        {1, 2, 3},
        {5, 6, 0},
        {7, 8, 4}
    };
    // Solvable Final configuration
    // Value 0 is used for empty space
    int final[N][N] =
    {
        {1, 2, 3},
        {5, 8, 6},
        {0, 7, 4}
    };
    // Blank tile coordinates in initial
    // configuration
    int x = 1, y = 2;
    solve(initial, x, y, final);
    return 0;
}"
```

Output

```
1 2 3
5 6 0
7 8 4

1 2 3
5 0 6
7 8 4

1 2 3
5 8 6
7 0 4

1 2 3
5 8 6
0 7 4
```

Case Study 4: Bouncing Ball

To save the file, either click on it or right-click it or pick Save Link As; you can insert any missing code in the Python file. The software with the extension .py should be downloaded. Double-click the file to open. You can open the Canopy Python framework and display the code in the editor window. This is what you need to see if you downloaded and opened the file successfully.

Consider the various tasks, their statements and the overall software structure. The key aspects we will work with are the keys for upgrading and developing; see Figure CS.9. This is the way the software runs.

The evolving function is the entry point of the software. The first rpm, ball height and time increase is taken by this method.

Under these initial conditions, the plotting setting is calculated based on the original position and speed of the ball (in two dimensions) and the main loop is initiated, which pushes the ball forward in time.

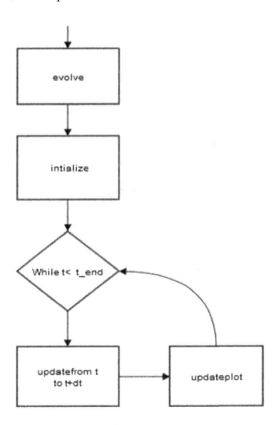

FIGURE CS.9

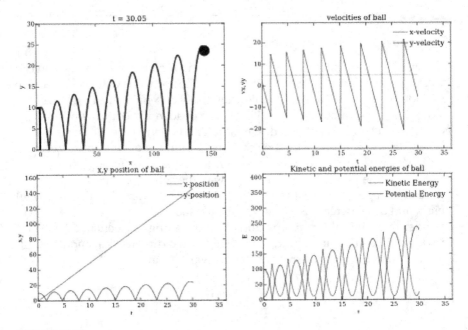

FIGURE CS.10

You can see a window of four plots if you filled in the missing code right.

- An animation of the ball is seen on the top-left map.
- The x and y speeds of the ball show in the top-right drawing through-out time.
- The lower left graph displays over time the ball's x and y positions.
- The lower right illustration displays the bouncing ball's movement and probable intensity over time.

Will the ball have something odd or unphysical when it bounces? Look at the energy charts as well as the ball bouncing. What should occur? What should happen? Your story should look like that at the end of the simulation; see Figure CS.10,

Code

```
"import pygame
import random
# Define some colors
BLACK = (0, 0, 0)
WHITE = (255, 255, 255)
```

```
SCREEN_WIDTH = 700
SCREEN_HEIGHT = 500
BALL_SIZE = 25
class Ball:
    """
    Class to keep track of a ball's location and vector.
    """
    def __init__(self):
        self.x = 0
        self.y = 0
        self.change_x = 0
        self.change_y = 0
def make_ball():
    """
    Function to make a new, random ball.
    """
    ball = Ball()
    # Starting position of the ball.
    # Take into account the ball size so we don't spawn on the
    edge.
    ball.x = random.randrange(BALL_SIZE, SCREEN_WIDTH - BALL_SIZE)
    ball.y = random.randrange(BALL_SIZE, SCREEN_HEIGHT
    - BALL_SIZE)
    # Speed and direction of rectangle
    ball.change_x = random.randrange(-2, 3)
    ball.change_y = random.randrange(-2, 3)
    return ball
def main():
    """
    This is our main program.
    """
    pygame.init()
    # Set the height and width of the screen
    size = [SCREEN_WIDTH, SCREEN_HEIGHT]
    screen = pygame.display.set_mode(size)
    pygame.display.set_caption("Bouncing Balls")
    # Loop until the user clicks the close button.
    done = False
    # Used to manage how fast the screen updates
    clock = pygame.time.Clock()
    ball_list = []
    ball = make_ball()
    ball_list.append(ball)
    # -------- Main Program Loop -----------
    while not done:
        # --- Event Processing
        for event in pygame.event.get():
            if event.type == pygame.QUIT:
```

```
done = True
elif event.type == pygame.KEYDOWN:
# Space bar! Spawn a new ball.
if event.key == pygame.K_SPACE:
ball = make_ball()
ball_list.append(ball)
# --- Logic
for ball in ball_list:
# Move the ball's center
ball.x += ball.change_x
ball.y += ball.change_y
# Bounce the ball if needed
if ball.y > SCREEN_HEIGHT - BALL_SIZE or ball.y < BALL_SIZE:
ball.change_y *= -1
if ball.x > SCREEN_WIDTH - BALL_SIZE or ball.x < BALL_SIZE:
ball.change_x *= -1
# --- Drawing
# Set the screen background
screen.fill(BLACK)
# Draw the balls
for ball in ball_list:
pygame.draw.circle(screen, WHITE, [ball.x, ball.y], BALL_SIZE)
# --- Wrap-up
# Limit to 60 frames per second
clock.tick(60)
# Go ahead and update the screen with what we've drawn.
pygame.display.flip()
# Close everything down
pygame.quit()
if __name__ == "__main__":
main()"
```

Case Study 5: Animations

This software includes a variety of blocks that bounce off the window edges. The blocks are colored and weighted differently and only move diagonally. We push the blocks on each iteration via the game loop in order to move them (thus making them look like they are moving). This looks as if the blocks pass across the screen.

```
"import pygame, sys, time
from pygame.locals import *
# set up pygame
pygame.init()
```

```
# set up the window
WINDOWWIDTH = 400
WINDOWHEIGHT = 400
windowSurface = pygame.display.set_mode((WINDOWWIDTH,
WINDOWHEIGHT), 0, 32)
pygame.display.set_caption('Animation')
# set up direction variables
DOWNLEFT = 1
DOWNRIGHT = 3
UPLEFT = 7
UPRIGHT = 9
MOVESPEED = 4
# set up the colors
BLACK = (0, 0, 0)
RED = (255, 0, 0)
GREEN = (0, 255, 0)
BLUE = (0, 0, 255)
# set up the block data structure
b1 = {'rect':pygame.Rect(300, 80, 50, 100), 'color':RED,
'dir':UPRIGHT}
b2 = {'rect':pygame.Rect(200, 200, 20, 20), 'color':GREEN,
'dir':UPLEFT}
b3 = {'rect':pygame.Rect(100, 150, 60, 60), 'color':BLUE,
'dir':DOWNLEFT}
blocks = [b1, b2, b3]
# run the game loop
while True:
   # check for the QUIT event
for event in pygame.event.get():
if event.type == QUIT:
        pygame.quit()
        sys.exit()
   # draw the black background onto the surface
   windowSurface.fill(BLACK)
    for b in blocks:
        # move the block data structure
    if b['dir'] == DOWNLEFT:
            b['rect'].left -= MOVESPEED
            b['rect'].top += MOVESPEED
        if b['dir'] == DOWNRIGHT:
            b['rect'].left += MOVESPEED   b['rect'].top +=
            MOVESPEED
        if b['dir'] == UPLEFT:
            b['rect'].left -= MOVESPEED
            b['rect'].top -= MOVESPEED
        if b['dir'] == UPRIGHT:
            b['rect'].left += MOVESPEED
            b['rect'].top -= MOVESPEED
```

```
            # check if the block has moved out of the window
        if b['rect'].top < 0:
            # block has moved past the top
            if b['dir'] == UPLEFT:
                b['dir'] = DOWNLEFT
            if b['dir'] == UPRIGHT:
                b['dir'] = DOWNRIGHT
        if b['rect'].bottom > WINDOWHEIGHT:
            # block has moved past the bottom
            if b['dir'] == DOWNLEFT:
                b['dir'] = UPLEFT
            if b['dir'] == DOWNRIGHT:
                b['dir'] = UPRIGHT
        if b['rect'].left < 0:
            # block has moved past the left side
            if b['dir'] == DOWNLEFT:
                b['dir'] = DOWNRIGHT
            if b['dir'] == UPLEFT:
                b['dir'] = UPRIGHT
        if b['rect'].right > WINDOWWIDTH:
            # block has moved past the right side
            if b['dir'] == DOWNRIGHT:
                b['dir'] = DOWNLEFT
            if b['dir'] == UPRIGHT:
                b['dir'] = UPLEFT
        # draw the block onto the surface
        pygame.draw.rect(windowSurface, b['color'], b['rect'])
    # draw the window onto the screen
    pygame.display.update()
    time.sleep(0.02)"
```

How the Animation Program Works

We have three colored blocks in this software which shift and knock off the walls. We first have to worry about how we want the blocks to pass; see Figure CS.11.

In one of four diagonal ways each block moves. A block can bounce off its side as it reaches the side of the window and travel in a new diagonal direction. As Figure CS.11 indicates, the blocks will bounce.

The new way a block goes after it recovers depends on two things: the way it was going before it rebounded, and which wall it rebounded from. A block will bounce in a total of eight ways: two separate ways for each of the four walls.

For instance, we would like a new path for the block to be up and right, if a block travels down and right to bounce off the bottom edge of the window.

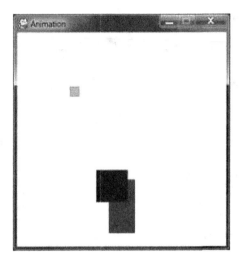

FIGURE CS.11
An altered screenshot of the animation program.

We may display blocks with a Rect object representing the block's location and height, a tuple of three integral elements to represent the block's color, and an integer representing one of the four diagonally moving positions in the block.

Change the location X and Y of the block in the object Rect for each iteration of the game loop. Draw all the blocks at their current location on the screen in each iteration. As the execution of the program is running through the game loop, the blocks are slowly moving and bouncing individually across the frame; see Figure CS.12.

Case Study 6: Shortest Path

In this graph you can find the shortest paths in the diagram, given a line and a source vertex.

The algorithm of Dijkstra is very similar to the minimal spanning tree (MST) Prim's algorithm. Like Prim's MST, we build a shortest path tree (SPT) with a root source. We have two sets, one set includes vertices in the SPT and a second set includes vertices that do not yet exist in the nearest neighbor tree. At each step of the algorithm, we have a vertex in the other set and the distance from the source is minimal.

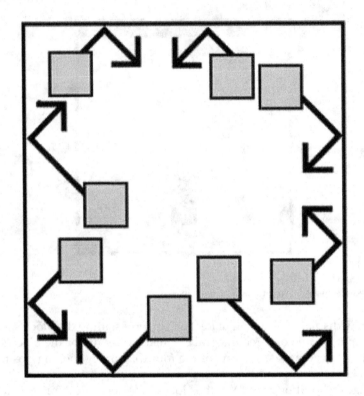

FIGURE CS.12

The elevated status used to determine the shortest path from a single source in the Dijkstra algorithm are:

1. Create a set sptSet (SPT set) that records vertices embedded in a smaller path tree, which measures and completes the minimum distance from the source. This set is empty in the beginning.

2. Give all the vertices of the input diagram a distance weight. Initialize Infinity for all distance quantities. Assign the source vertex distance value to 0 so it is chosen first.

3. Although not all vertices are used in sptSet:
   - Select a u vertex not present in sptSet with the least distance value.
   - U to sptSet included.
   - Check all neighboring vertical u distance values. Use a neighboring vertical to change distance values. If the sum of a distance value of u (from the source) and a weight of edge u-v for each neighboring vertex v is less than the distance value of v, then change the maximum distance v.

Code

```
"# Python program for Dijkstra's single
# source shortest path algorithm. The program is
# for adjacency matrix representation of the graph

# Library for INT_MAX
import sys

class Graph():

    def __init__(self, vertices):
        self.V = vertices
        self.graph = [[0 for column in range(vertices)]
                    for row in range(vertices)]

    def printSolution(self, dist):
        print ("Vertex tDistance from Source")
        for node in range(self.V):
            print (node, "t", dist[node])

    # A utility function to find the vertex with
    # minimum distance value, from the set of vertices
    # not yet included in shortest path tree
    def minDistance(self, dist, sptSet):

        # Initilaize minimum distance for next node
        min = sys.maxsize

        # Search not nearest vertex not in the
        # shortest path tree
        for v in range(self.V):
            if dist[v] < min and sptSet[v] == False:
                min = dist[v]
                min_index = v

        return min_index

    # Funtion that implements Dijkstra's single source
    # shortest path algorihm for a graph represented
    # using adjacency matrix representation
    def dijkstra(self, src):

        dist = [sys.maxsize] * self.V
        dist[src] = 0
        sptSet = [False] * self.V

        for cout in range(self.V):

            # Pick the minimum distance vertex from
            # the set of vertices not yet processed.
            # u is always equal to src in first iteration
            u = self.minDistance(dist, sptSet)

            # Put the minimum distance vertex in the
```

```
        # shotest path tree
        sptSet[u] = True

        # Update dist value of the adjacent vertices
        # of the picked vertex only if the current
        # distance is greater than new distance and
        # the vertex in not in the shotest path tree
        for v in range(self.V):
            if self.graph[u][v] > 0 and \
                sptSet[v] == False and \
                dist[v] > dist[u] + self.graph[u][v]:
                dist[v] = dist[u] + self.graph[u][v]

    self.printSolution(dist)

# Driver program
g = Graph(9)
g.graph = [[0, 4, 0, 0, 0, 0, 0, 8, 0],
        [4, 0, 8, 0, 0, 0, 0, 11, 0],
        [0, 8, 0, 7, 0, 4, 0, 0, 2],
        [0, 0, 7, 0, 9, 14, 0, 0, 0],
        [0, 0, 0, 9, 0, 10, 0, 0, 0],
        [0, 0, 4, 14, 10, 0, 2, 0, 0],
        [0, 0, 0, 0, 0, 2, 0, 1, 6],
        [8, 11, 0, 0, 0, 0, 1, 0, 7],
        [0, 0, 2, 0, 0, 0, 6, 7, 0]
        ];

g.dijkstra(0);

Vertex tDistance from Source
0 t 0
1 t 4
2 t 12
3 t 19
4 t 21
5 t 11
6 t 9
7 t 8
8 t 14"
```

Case Study 7: Tree Traversal Algorithm

Binary Tree and Its Traversal Using Python

A binary tree is the tree where only two children can be a node and no more than two. Crossing means visiting the binary tree nodes. Three kinds of crossing exist; see Figure CS.13.

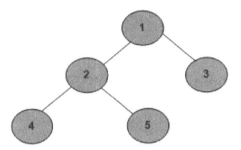

FIGURE CS.13

In-order traversal

In order traversal means visiting the left first, then the root and then the right.

So the traversal of the above tree would be 4 2 5 1 3.

Pre-order traversal

In this traversal we first visit the root, then the left and then the right.

It will be something like this 1 2 4 5 3.

Post-order traversal

Here we first visit the left, then the right and then the root.

It will be something like this 4 5 2 3 1.

Code

```
"class Node:
    def __init__(self,data):
        self.left = None
        self.right = None
        self.data = data
def inOrder(root):
    if root:
        inOrder(root.left)
        print (root.data)
        inOrder(root.right)
def preOrder(root):
    if root:
        print (root.data)
        preOrder(root.left)
        preOrder(root.right)
def postOrder(root):
    if root:
```

```
        postOrder(root.left)
        postOrder(root.right)
        print (root.data)

#making the tree
root = Node(1)
root.left = Node(2)
root.right = Node(3)
root.left.left = Node(4)
root.left.right = Node(5)

print inOrder(root)
#4 2 5 1 3
print preOrder(root)
#1 2 4 5 3
print postOrder(root)
#4 5 2 3 ?"
```

References

[1] D. Arnold, M.A. Bond Chilvers and R. Taylor, Hector: Distributed objects in Python, In *Proceedings of the 4th International PythonConference*, Australia, 1996.

[2] D. Ascher, P.F. Dubois, K. Hinsen, J. Hugunin and T. Oliphant Numerical Python, Technical report, Lawrence Livermore National Lab., CA, 2001. http://www.pfdubois.com/numpy/numpy.pdf

[3] D. Beazley *Python Essential Reference*, 2nd edition. New Riders Publishing, Indianapolis, 2001.

[4] D. Beazley et al., Swig 1.3 Development Documentation, Technical report, 2004. http://www.swig.org/doc.html

[5] Guido van Rossum and F.L. Drake Extending and embedding the Python interpreter, 2006. http://docs.python.org/ext/ext.html

[6] Blitz++software. http://www.oonumerics.org/blitz/

[7] W.L. Briggs *A Multigrid Tutorial*. SIAM Books, Philadelphia, PA, 1987.

[8] A.M. Bruaset *A Survey of Preconditioned Iterative Methods*. Addison-Wesley Pitman, Reading, MA, 1995.

[9] A. Grama, A. Gupta, G. Karypis and V. Kumar *Introduction to Parallel Computing*, 2nd edition. Addison–Wesley, Reading, MA, 2003.

[10] W. Gropp, E. Lusk and A. Skjellum *Using MPI – Portable Parallel Programming with the Message-Passing Interface*, 2nd edition. MIT Press, Cambridge, MA, 1999.

[11] K. Jackson pyGlobus: A Python interface to the Globus toolkit. *Concurrency and Computation: Practice and Experience* 14, 1075–1084, 2002.

[12] H.P. Langtangen Python scripting for computational science, *Texts in Computational Science and Engineering* (Vol. 3). Springer, Berlin Heidelberg, 2004.

[13] Matlab code vectorization guide, 2007. http://www.mathworks.com/support/tech-notes/1100/1109.html

[14] S. Montanaro Python performance tips. http://manatee.mojam.com/~skip/python/fastpython.html

[15] C. Ramu and C. Gemuend CGIMODEL: CGI programming made easy with Python. *Linux Journal* 6, 235–250, 2000.

[16] M.F. Sanner Python: A programming language for software integration and development. *Journal of Molecular Graphics and Modelling* 17(1), 57–61, 1999.

[17] Guido van Rossum A switch/case statement, June 2007. http://www.python.org/dev/peps/pep-3103/

[18] P. Peterson F2PY: A tool for connecting Fortran and Python programs. *International Journal of Computational Science and Engineering* 4(4), 296, 2009. http://dx.doi.org/10.1504/IJCSE.2009.02916521.Pellerin J 2009 'nose'. https://nose.readthe-docs.org/en/latest/

[19] Sphinx: Python documentation generator, 2015. http://sphinx-doc.org/

[20] numpydoc – Numpy's sphinx extensions, 2015. https://github.com/numpy/numpydoc

[21] F. Pedregosa, G. Varoquaux, A. Gramfort, V. Michel, B. Thirion, O. Grisel, M. Blondel, P. Prettenhofer, R. Weiss, V. Dubourg, J. Vanderplas, A. Passos, D. Cournapeau, M. Brucher, M. Perrot and É. Duchesnay Scikit-learn: Machine learning in Python. *Journal of Machine Learning Research* 12, 2825–2830, 2011.

[22] Fernando Perez and Brian E. Granger IPython: A system for interactive scientific computing. *Computing in Science & Engineering* 9(3), 21–29, May/June 2007. https://doi.org/10.1109/MCSE.2007.53.

[23] S. Behnel, R. Bradshaw, D.S. Seljebotn, G. Ewing et al. C-Extensions for Python. http://www.cython.org

[24] J. Bergstra Optimized symbolic expressions and GPU metaprogramming with Theano, In *Proceedings of the 9th Python in Science Conference (SciPy2010)*, Austin, Texas, June 2010.

[25] D. Cooke, F. Alted, T. Hochberg and G. Thalhammer num expr http://code.google.com/p/numexpr/ NumPy's documentation is maintained using a WikiPedia-like community forum, http://docs.scipy.org/. Discussions take place on the project mailing list (http://www.scipy.org/Mailing_Lists).

[26] C. Fuhrer, J.E. Solem and O. Verdier *Scientific Computing with Python 3*. Packt Publishing Ltd, Birmingham, 2016.

[27] D. Harms and K. McDonald, *The Quick Python Book*. Manning, Greenwich, 1999.

[28] S. Šandi, T. Popović and B. Krstajić Python implementation of IEEE C37. 118 communication protocol. *ETF Journal of Electrical Engineering* 21(1), 108–117, 2015.

[29] Olle Bälter and Duane A. Bailey Enjoying Python, processing, and Java in CS1. *ACM in Roads* 4(2010), 28–32, 2010. https://doi.org/10.1145/1869746.1869758

[30] Stephen H. Edwards, Daniel S. Tilden and Anthony Allevato Pythy: Improving the introductory Python programming experience. In *Proceedings of the 45th ACM Technical Symposiumon Computer Science Education (SIGCSE '14)* (pp. 641–646), ACM, New York, NY, 2014. https://doi.org/10.1145/2538862.2538977

[31] Richard J. Enbody, William F. Punch and Mark McCullen Python CS1 as preparation for C++ CS2, In *Proceedings of the 40th ACM Technical Symposiumon Computer Science Education (SIGCSE '09)* (pp. 116–120), ACM, New York, NY, 2009. https://doi.org/10.1145/1508865.1508907

[32] Thomas R. Etherington Teaching introductory GIS programming to geographers using an open source Python approach. *Journal of Geography in Higher Education* 40(1), 117–130, 2016. https://doi.org/10.1080/03098265.2015.1086981

[33] Geela Venise Firmalo Fabic, Antonija Mitrovic, and Kourosh Neshatian Investigating the effectiveness of menu-based self-explanation prompts in a mobile Python tutor. In Elisabeth André, Ryan Baker, Xiangen Hu, Ma. Mercedes T. Rodrigo and Benedict du Boulay (Eds.) *Artificial Intelligence in Education*. Springer International Publishing, Cham, 498–501, 2017. https://doi.org/10.1007/978-3-319-61425-0_49

[34] Ambikesh Jayal, Stasha Lauria, Allan Tucker and Stephen Swift Python for teaching introductory programming: A quantitative evaluation. *ITALICS Innovations in Teachingand Learning in Information and Computer Sciences* 10(1), 86–90, 2011.

[35] M. Poole Extending the design of a blocks-based Python environment to support complex types, In *IEEE Blocks and Beyond Workshop (B&B '17) IEEE* (pp. 1–7), United Kingdom, 2017. https://doi.org/10.1109/BLOCKS.2017.8120400

[36] Kelly Rivers and Kenneth R. Koedinger Data-driven hint generation invast solution spaces: A self-improving Python programming tutor. *International Journal of Artificial Intelligence in Education* 27(1), 37–64, 2015. https://doi.org/10.1007/s40593-015-0070-z

[37] Hong Wang Teaching CS1 with Python GUI game programming, In *AIP Conference Proceedings* (pp. 253–260), Vol. 1247, AIP, San Francisco CA, 2010. https://doi.org/10.1063/1.3460234

[38] M. Korzen and S. Jaroszewicz Pacal: A Python package for arithmetic computations with random variables. *Journal of Statistical Software* 57, 1, 2014.

[39] A. Garrett inspyred: Python library for bio-inspired computational intelligence, 2019, https://github.com/aarongarrett/inspyred (Accessed 6 May 2019).

[40] D. Hadka Platypus: Multiobjective optimization in Python, https://platypus.readthedocs.io (Accessed 16 May 2019).

[41] E. Zitzler and L. Thiele Multiobjective optimization using evolutionary algorithms – A comparative case study, In *Proceedings of the 5th International Conference on Parallel Problem Solving fromNature* (pp. 292–304), Springer-Verlag, London, UK, 1998. http://dl.acm.org/citation.cfm?id=645824.668610

[42] Y.S. Tan and N.M. Fraser The modified star graph and thepetal diagram: Two new visual aids for discrete alternative multicrite-ria decision making. *Journal of Multi-Criteria Decision Analysis* 7, 20–33, 1998.

[43] Y. Tian, R. Cheng, X. Zhang and Y. Jin PlatEMO: A MATLAB platform for evolutionary multi-objective optimization. *IEEE Computational Intelligence Magazine* 12, 73–87, 2017.

[44] L. Rachmawati and D. Srinivasan Multi-objective evolutionary algorithm with controllable focus on the knees of the pareto front. *Evolutionary Computation* 13, 810–824, 2009. https://doi.org/10.1109/TEVC.2009.2017515.

[45] D. Maclaurin, D. Duvenaud and R.P. Adams Autograd: Effortless gradients in numpy, In *ICML 2015 Auto ML Workshop*, 2015. https://indico.lal.in2p3.fr/event/2914/session/1/contribution/6/3/material/paper/0.pdf; https://github.com/HIPS/autograd

[46] M.D. McKay, R.J. Beckman and W.J. Conover A comparison of three methods for selecting values of input variables in the analysis of output from a computer code. *Technimetrics* 42, 55–61, 2000. http://dx.doi.org/10.2307/1271432

[47] Alfred V. Aho, John E. Hopcroft and Jeffrey D. Ullman *Data Structures and Algorithms*. Addison-Wesley, Reading, MA, 1983.

[48] John L. Bentley Programming pearls: How to sort. *Communications of the ACM* 27(3), 287–291, March 1984.

[49] John L. Bentley Programming pearls: The back of the envelope. *Communications of the ACM* 27(3), 180–184, March 1984.

[50] John L. Bentley Programming pearls: Thanks, heaps. *Communications of the ACM* 28(3), 245–250, March 1985.

[51] John L. Bentley *Programming Pearls*. Addison-Wesley, Reading, MA, 1986.

[52] John L. Bentley Programming pearls: The envelope is back. *Communications of the ACM* 29(3), 176–182, March 1986.

[53] Timothy Budd *Classic Data Structures*. Addison-Wesley, Boston, 2001.

[54] Thomas H. Cormen, Charles E. Leiserson and Ronald L. Rivest *Introduction to Algorithms*. MIT Press, Cambridge, MA, 1990.

[55] Nell B. Dale *C++ Plus Data Strucutres*. Jones and Bartlett, Sudbury, MA, 1999.

[56] Nell B. Dale and Susan C. Lilly *Pascal Plus Data Structures, Algorithms and Advanced Programming Sua*. Houghton Mifflin Co., Boston, MA, 1995.

[57] Donald E. Knuth *Sorting and Searching, Volume 3 of The Art of Computer Programming*, 2nd edition. Addison-Wesley, Reading, MA, 1981.

[58] G. Van Rossum and J. de Boer Interactively testing remote servers using the Python programming language. *CWI Quarterly* 4, 283–303, 1991.

[59] P.H. Chou Algorithm education in Python, In *10th International Python Conference*, 4–7 February 2002, Alexandria, Virginia, 2002. http://www.python10.org/p10-papers/index.htm (Accessed 4 October 2007).

[60] T. Hamelryck and B. Manderick PDB file parser and structure class implemented in Python. *Bioinformatics* 19, 2308–2310, 2003.

[61] M. Hammond Python programming on Win32 using PythonWin. In M. Hammond and A. Robinson (Eds.) *Python Programming on Win32*. O'Reilly Network, USA, 2007. http://www.onlamp.com/pub/a/python/excerpts/chpt20/pythonwin.html (Accessed 4 October 2007).

[62] F.H. Perez, B.E. Granger Ipython: A system for interactive scientific computing. *Computing in Science and Engineering* 9, 21–29, NumPy. Trelgol Publishing, 2007. http://numpy.scipy.org/ (Accessed 4 October 2007).

[63] Brian Beck Readable switch construction without lambdas or dictionaries, April 2005. http://aspn.activestate.com/ASPN/Cookbook/Python/Recipe/410692

[64] J. He, W. Shen, P. Divakaruni, L. Wynter and R. Lawrence Improving traffic prediction with tweet semantics, In *Proceedings of the Twenty-Third International Joint Conference on Artificial Intelligence* (pp. 1387–1393), IBM Researchers, August 3–9, 2013.

[65] A. Agarwal, B. Xie, I Vovsha, O. Rambow and R. Passonneau Sentiment analysis of Twitter data, In *The Proceedings of Workshop on Language in Social Media*, ACL, 2011.

[66] S. Kumar, F. Morstatter and H. Liu *Twitter Data Analytics*. Springer Book, New York, 2013.

[67] A. Mittal and A. Goel *Stock Prediction Using Twitter Sentiment Analysis*. Stanford University, Stanford, CA, 2011.

[68] D. Ediger, K. Jiang, J. Riedy and D.A. Bader Massive social network analysis: Mining twitter for social good, In *39th International Conference on Parallel Processing* (pp. 583–593), UK, 2010.

[69] Nikita Pilnenskiy and Ivan Smetannikov *Modern Implementations of Feature Selection Algorithms and Their Perspectives*. ITMO University, St.Petersburg, Russia, 2019.

[70] R. Tohid, Bibek Wagle, Shahrzad Shirzad, Patrick Diehl, Adrian Serio, Alireza Kheirkhahan, Parsa Amini, Katy Williamst, Kate Isaacst, Kevin Huck, Steven Brandt and Hartmut Kaiser *Asynchronous Execution of Python Code on Task-based Runtime Systems*. Louisiana State University, University of Arizona, University of Oregon, 2018.

[71] Abhinav Nagpal and Goldie Gabrani *Python for Data Analytics, Scientific and Technical Applications*, 2019.

[72] A. Watson, D.S.V. Babu and S. Ray Sanzu: A data science benchmark, In *2017 IEEE International Conference on Big Data (Big Data)* (pp. 263–272), Australia, 2017.

[73] I. Stančin and A. Jović An overview and comparison of free Python libraries for data mining and big data analysis, In *2019 42nd International Convention on Information and Communication Technology, Electronics and Microelectronics (MIPRO)* (pp. 977–982), 2019.

[74] Mark Guzdial *Introduction to Computing and Programming in Python: A Multimedia Approach*. Prentice Hall, Upper Saddle River, NJ, 2005.

[75] David Harel *Algorithmics: The Spirit of Computing*. Addison Wesley, Harlow, 2004.

[76] Trevor Hastie, Robert Tibshirani and Jerome Friedman *The Elements of Statistical Learning: Data Mining, Inference, and Prediction*, 2nd edition. Springer, New York, 2009.

[77] Marti Hearst Automatic acquisition of hyponyms from large text corpora, In *Proceedings of the 14th Conference on Computational Linguistics (COLING)* (pp. 539–545), 1992.

[78] Irene Heim and Angelika Kratzer *Semantics in Generative Grammar*. Blackwell, Oxford, 1998.

[79] Lynette Hirschman, Alexander Yeh, Christian Blaschke and Alfonso Valencia Overview of biocreative: Critical assessment of information extraction for biology. *BMC Bioinformatics* 6(Supplement 1), May 2005.

[80] Wilfred Hodges *Logic*. Penguin Books, Harmondsworth, 1977. [Huddleston and Pullum, 2002] Rodney D. Huddleston and Geoffrey K. Pullum. *TheCambridge Grammar of the English Language*. Cambridge University Press, 2002.

[81] Andrew Hunt and David Thomas *The Pragmatic Programmer: From Journeyman to Master*. Addison Wesley, Reading, MA, 2000.

[82] Nitin Indurkhya and Fred Damerau editors. *Hand-book of Natural Language Processing*, 2nd edition. CRC Press, Taylor and Francis Group, Boca Raton, 2010.

[83] Ray Jackend off *X-Syntax: A Study of Phrase Structure. Number 2 in Linguistic Inquiry Monograph*. MIT Press, Cambridge, MA, 1977.

[84] Mark Johnson *Attribute Value Logic and Theory of Grammar. CSLI Lecture Notes Series*. University of Chicago Press, Chicago, IL, 1988.

[85] Daniel Jurafsky and James H. Martin *Speech and Language Processing*, 2nd edition. Prentice Hall, Upper Saddle River, NJ; London, 2008.

[86] M.E. Caspersen and M. Kölling A novice's process of object-oriented programming, In *Companion to the 21st ACM Sigplan Symposium on Object-oriented Programming Systems, Languages, and Applications* (pp. 892–900), ACM, New York, NY, 2006. https://doi.org/10.1145/1176617.1176741

[87] C. Dierbach Python as a first programming language. *Journal of Computing Sciences in Colleges* 29(6), 153–154, June 2014. http://dl.acm.org/citation.cfm?id=2602724.2602754

[88] S.H. Edwards, D.S. Tilden and A. Allevato Pythy: Improving the introductory Python programming experience, In *Proceedings of the 45th ACM Technical Symposium on Computer Science Education* (pp. 641–646), ACM, New York, NY, 2014. https://doi.org/10.1145/2538862.2538977

[89] R.J. Enbody and W.F. Punch Performance of Python CS1 students in mid-level non-python CS courses, In *Proceedings of the 41st ACM Technical Symposium on Computer Science Education* (pp. 520–523), ACM, New York, NY, 2010. https://doi.org/10.1145/1734263.1734437

[90] K. Ericsson and H. Simon *Protocol Analysis: Verbal Reports as Data*, revised edition. MIT Press, Cambridge, MA, 1993.

[91] A.E. Fleury Parameter passing: The rules the students construct, In *Proceedings of the Twenty-second Sigcse Technical Symposium on Computer Science Education* (pp. 283–286), ACM, New York, NY, 1991.

[92] M.H. Goldwasser and D. Letscher Teaching an object-oriented CS1 – With python. *Acm Sigcse Bulletin* 40, 42–46, 2008.

[93] S. Holland, R. Griffiths and M. Woodman Avoiding object misconceptions. *SIGCSE Bull* 29(1), 131–134, 1997.

[94] N. Liberman, C. Beeri and Y. Ben-David Kolikant Difficulties in learning inheritance and polymorphism. *ACM Transactions on Computing Education (TOCE)* 11(1), 4, 2011.

[95] R. Lister, A. Berglund, T. Clear, J. Bergin, K. Garvin-Doxas, B. Hanks, . . .J.L. Whalley Research perspectives on the objects-early debate, In Working group reports on *ITICSE on Innovation and Technology in Computer Science Education* (pp. 146–165), ACM, New York, NY, 2006. https://doi.org/10.1145/1189215.1189183

[96] A.D. Moore *Python GUI Programming with Tkinter*. Packt Publishing Ltd., Birmingham, 2018.

[97] B. Chaudhary *Tkinter GUI Application Development Blueprints: Build Nine Projects by Working with Widgets, Geometry Management, Event Handling, and More*. Packt Publishing Ltd., Birmingham, 2018.

[98] D. Love *Tkinter GUI Programming by Example*. Packt Publishing Ltd., Birmingham, 2018.

[99] J.E. Grayson *Python and Tkinter Programming*. Manning, Greenwich, 2000.

[100] A. Rodas de Paz *Tkinter GUI Application Development Cook Book*. Packt Publishing Ltd., Birmingham, 2018.

[101] M. Roseman *Modern Tkinter for Busy Python Developers: Quickly Learn to Create Great Looking User Interfaces for Windows, Mac and Linux using Python's Standard GUI Toolkit*. Late Afternoon Press, 2012

[102] Young Douglas *The X Window System: Programming and Applications with Xt, OSF/Motif*, 2nd edition. Prentice Hall, Englewood Cliffs, ISBN: 0-13123-803-5, 1994.

[103] Flynt Clifton *Tcl/Tk for Real Programmers*. Academic Press (AP Professional), San Diego, ISBN 0-12261-205-1, 1998.

[104] Foster-Johnson Eric *Graphical Applications with Tcl and Tk*, 2nd edition. M&T Books, New York, ISBN 1-55851-569-0, 1997.

[105] Harrison Mark and Michael J. McLennan *Effective Tcl/Tk Programming: Writing Better Programs in Tcl and Tk*. Addison Wesley Longman, Reading, MA, ISBN: 0-20163-474-0, 1997.

[106] Tkinter Wiki, http://tkinter.unpythonic.net/wiki/ (Accessed 4 October 2007).

[107] J.J. Horning, H.C. Lauer, P.M. Melliar-Smith and B. Randell A program structure for error detection and recovery, In *Proceedings of Conference on Operating Systems, IRIA* (pp. 177–193), Serbia, 1974.

[108] B. Randell System structure for software fault tolerance. *IEEE Transactions on Software Engineering* 1(1), 220–232, June 1975.

[109] J.B. Goodenough Exception handling: Issues and a proposed notation. *Communications of the ACM* 18(12), 683–696, ACM Press, December 1975.

[110] F. Cristian Exception handling and software fault tolerance, In *Proceedings of FTCS-25*, 3, IEEE (reprinted from FTCS-IO 1980, 97–103), Japan, 1996.

[111] A. Garcia, C. Rubira, A. Romanovsky and J. Xu A comparative study of exception handling mechanisms for building dependable object-oriented software. *Journal of Systems and Software* 2, 197–222, November 2001.

[112] Guido van Rossum and Fred L. Drake *An Introduction to Python – The Python Tutorial*. Network Theory Ltd, Bristol, 2006.

[113] Guido van Rossum and Fred L. Drake *The Python Language Reference Manual*. Network Theory Ltd, Bristol, 2006.

[114] Jesus Mena *Investigative Data Mining for Security and Criminal Detection*. Butterworth-Heinemann, Amsterdam; Boston, MA, 2003.

[115] Kelvin Chan and Jay Leibowitz The synergy of social network analysis and knowledge mapping: A case study. *International Journal of Management and Decision Making* 7(1), 19, 2006.

[116] Kate Ehrlich and Inga Carboni working paper, IBM Watson Research Center. 4. S. Wasserman and K. Faust Social Network Analysis: Methods and Applications. Cambridge University Press, Cambridge, (1994).

Index

Printed in the United States
by Baker & Taylor Publisher Services